■ FROM HAND TO MOUTH ■

Michael C. Corballis

■ FROM HAND TO MOUTH ■

THE ORIGINS OF LANGUAGE

PRINCETON UNIVERSITY PRESS

PRINCETON AND OXFORD

Copyright © 2002 by Princeton University Press

Published by Princeton University Press, 41 William Street, Princeton, New Jersey
08540

In the United Kingdom: Princeton University Press, 3 Market Place, Woodstock,
Oxfordshire OX20 1SY

Library of Congress Cataloging-in-Publication Data

Corballis, Michael C.

From hand to mouth : the origins of language / Michael C. Corballis.

p.   cm.

Includes bibliographical references and index.

ISBN 0-691-08803-9 (alk. paper)

1. Language and languages—Origin.   I. Title.

P116 .C67 2002

401—dc21   2001058003

British Library Cataloging-in-Publication Data is available

This book has been composed in ITC Garamond Light with Frutiger.

Printed on acid-free paper. ∞

www.pup.princeton.edu

Printed in the United States of America

10 9 8 7 6 5 4 3 2 1

■ CONTENTS ■

■ PREFACE ■

When Tom and Elizabeth took the farm
　The bracken made their bed,
And *Quardle oodle ardle wardle doodle*
　The magpies said.

So wrote the New Zealand poet Denis Glover, in his poem *The Magpies*. I too was raised on a New Zealand farm and can remember the shrill, yet strangely melodious, sounds of the magpies that nested in the trees around the house. (Lest you wonder at the strangeness of these sounds, I should point out that these magpies differed from the North American variety; they were of Australian origin, and Australian sounds are often quite unlike any others.) I can also remember wondering what they were talking about.

Perhaps I shouldn't have wondered. One day, a magpie swooped and stabbed me on the top of the head—not a serious wound, but a warning that I should mind my own business. My father appeared shortly afterward with a shotgun and shot the offending bird. But in any event I now know that the *quardle oodle ardle* of the magpies has little resemblance to human speech, despite the Greek biographer Plutarch's account of a talking magpie that belonged to a barber in Rome. *Quardle schmardle.*

As I shall try to show in this book, human language has a complexity and creativity that is unmatched by any other form of animal communication, and probably depends on completely different principles. But if it is indeed true that there are no antecedents of language to be found elsewhere in the animal kingdom, then it may seem we have little upon which to build a theory of language evolution. In 1866 the Linguistic Society of Paris banned all discussion of language evolution. Quite apart from the problem of the lack of evidence, the topic may simply have been too hot. Just seven years earlier, Charles Darwin's *Origin of Species* had appeared and provoked enormous controversy, especially for the clear implication, however obliquely Darwin expressed it, that humans are descended from apes. If language is one thing we might call our own, it might be better not to examine it too closely, lest we discover that orangutans are talking about us behind our backs. Queen Victoria had already sounded a note of disapproval. On her first visit to the London Zoo in 1842, she is reported to have said "The Orang Outan is too wonderful. . . . He is frightful and painfully and disagreeably human."[1]

In his book *Almost Like a Whale*, designed as an update of Darwin's *Origin of Species*, the geneticist Steve Jones writes, "When it comes to what makes us unique, science can answer all the questions except the interesting ones."[2] For example, he suggests that we can never know the origins of the conscious mind: "We all have one, but as far as we can tell nothing else does. As a result, to speculate about its evolution is futile."[3] Perhaps language is in the same category.

But I think the Paris linguists were too severe, and Jones too pessimistic. I shall try to show in this book that we actually have quite a lot of evidence that bears on the evolution of language. Over the past seventy-five years, a great many fossil discoveries have provided increasingly detailed, if often controversial, accounts of the evolution of our species in the 5 or 6 million years since the line leading to modern humans split from the line leading to modern chimpanzees and bonobos. Molecular biologists have now added to the story. Those studying the behavior of present-day primates have given increasingly detailed accounts of behaviors that bear on language, even if they do not constitute language itself. Further evidence has come from fields as diverse as neuroscience, linguistics, anthropology, and developmental psychology. My aim in this book is to

[1] Quoted in Jones 2000, 431.
[2] Jones 2000, 432–33.
[3] Ibid., 433. Not everyone would agree that only humans have conscious minds.

weave a story about the evolution of language from threads drawn from a broad range of disciplines, in a way that I hope is accessible to all who are interested in the issue.

The main theme of the story is that language evolved, not from the vocal calls of our primate ancestors, but rather from their manual and facial *gestures*. Given that we are such a talkative species, this may seem perverse. Nevertheless, the idea is an old one, dating at least from the centuries of European expansionism, when traders found they could communicate more easily with the indigenous peoples they encountered by using manual gestures than they could using words. More formally, the gestural theory is often attributed to the eighteenth-century philosopher Condillac, although he had to disguise his theory in the form of a fable, since it was widely held at the time that language was so special it must have been bestowed on the human race by God. Although the gestural theory has been advocated many times since Condillac proposed it, it is probably still a minority view. You have been warned. Nevertheless, I invite you to watch more closely to what people do with their hands when they speak. You may perhaps agree with me that the transition of language from hand to mouth is still not fully complete.

■ ACKNOWLEDGMENTS ■

In her autobiography *Memories of a Catholic Girlhood*, the American novelist Mary McCarthy recounts how she decided, as a child, to affect to lose the Catholic faith. She was motivated initially more by the drama of it than by conviction, but the priest was called in to reason with her, and the more he tried to persuade her, the more she found counterarguments to support her stance.

In this book I am similarly indebted to those who have tried to talk me out of the view that language originated in manual gestures, as well as to those who have helped me find additional support. I owe thanks to the following, some of whom don't realize how much they have helped me, and some of whom would disagree with much of what I have to say: Giovanni Berlucchi, Ellen Bialystok, Derek Bickerton, Dick Byrne, Paul Corballis, Iain Davidson, Jill de Villiers, Merlin Donald, Russ Genet, Tom Givón, Russell Gray, Jim Hurford, Steve Keele, Tony Lambert, Stephen Lea, Andrew Lock, Carol Patterson, Anne Russon, Vince Sarich, the late William C. Stokoe, Tom Suddendorf, Sherman Wilcox, and the year 2000 graduate seminar in evolutionary psychology at the University of Auckland. Margaret Francis kindly helped me with the illustrations. I also extend special thanks to Vicky Wilson-Schwartz, who copyedited the manuscript, and to Sam

Elworthy, my editor at Princeton University Press. Sam's contribution was incalculable, and as a fellow countryman he was tolerant of my native quirks.

And special thanks too to Barbara Corballis, who put up with all this.

■ **FROM HAND TO MOUTH** ■

1 ■ What Is Language? ■

I am beguiled by the frivolous thought that we are descended, not from apes, but from birds. We humans have long sought features that are unique to our own species, with an especially keen eye for those that show us to be superior to others. Many special qualities differentiating us from our ape cousins have been proposed, but often, disconcertingly, these are found in our feathered friends as well. Like us, birds get around on two legs rather than four, at least when they're not flying (and some of them can't). Parrots, at least, have a consistent preference for picking things up with one foot, although in a mocking reversal of human handedness most of them prefer to use the left foot (most humans are right-handed and right-footed). Some birds prudently store food for the winter, and there is evidence that some of them can remember not only *where* they store food but also *when* they stored it, suggesting a kind of memory—known as episodic memory—that has been claimed as unique to our own species.[1] Birds make tools. They fly, albeit without purchasing airline tickets. They sing. And some of them talk.

Perhaps it is the last point that is the most interesting. Most birds far outperform mammals, including our immediate primate ancestors, in the variety and flexibility of the vocal sounds they make, and one can see (or hear) some striking parallels with

[1] The clever birds that do this are scrub jays, and their exploits are described in Clayton and Dickinson 1998.

human speech. The vocalizations of songbirds are complex and, like human speech, are controlled primarily by the left side of the brain.[2] Although birdsong is largely instinctive, birds can learn different dialects, and some of them can even learn arbitrary sequences of notes. In order to learn a particular song, the bird must hear it early on, while it is still in the nest, even though it does not produce the song until later. This crucial window of time is known as a *critical period*. The way people learn to speak also seems to depend on a critical period; that is, it seems to be impossible to learn to speak properly if we are not exposed to speech during childhood, and a second language learned after puberty is almost inevitably afflicted with a telltale accent. Some birds, like the parrot, can outdo humans in their ability to adapt their vocalizations, and not just by imitating human speech. The Australian lyre bird is said to be able to produce a near perfect imitation of the sound of a beer can being opened—which is perhaps the most commonly heard sound where humans congregate in that country.[3]

But of course birdsong differs in lots of ways from human speech. The ability of birds to imitate sounds probably has to do with the recognition of kin and the establishment and maintenance of territory but has nothing to do with conversation. Birds sing characteristic songs for much the same reason that nations of people fly characteristic flags or play national anthems. The remarkable ability of species like the mockingbird to imitate the songs of other birds has no doubt evolved also as a deceptive device to give the illusion of a territory filled with other birds, so that they may occupy that territory for themselves.[4]

Among most species of songbirds, it is only the males that vocalize, whereas women are said to be the more verbal members of our own species; we strong, silent chaps don't seem to have much to say. The vocalizations of birds, and indeed of other species, are mostly emotional, serving to signal aggression, to warn of danger, to advertise their sexual prowess, or to establish and maintain hierarchical social structures. Some of our own vocalizations serve similar, largely emo-

[2] For a review of this and other asymmetries in birds and other species, see Bradshaw and Rogers 1993.

[3] Although no longer, perhaps. Australia is now one of the great wine-producing countries, and wine has now largely replaced beer as the national drink. You can tell how many bottles of wine Australians have drunk by counting the number of corks dangling from their hats.

[4] Actually, among birds and other animals, the communications between nonkin (unrelated animals) are usually intended to deceive (e.g., Dawkins and Krebs 1978). As I jog along the beach, I am occasionally threatened by swooping, screeching seagulls and hope they are only kidding. Curiously, they stop if I walk instead of run, although I can scarcely tell the difference myself. It has been claimed that the deceptive quality of animal communication is a characteristic that distinguishes other species from ourselves, on the assumption that we don't usually lie, even to nonrelatives (Bingham 1999). I'm not so sure.

tional ends. We laugh, grunt, weep, shriek with fear, howl with rage, cry out in warning. But these noises, although important means of communication, are not *language*, as I explain below.

In any event, it would of course be irresponsible of me to claim any real kinship between humans and birds. There is a remote sense in which we *are* related to them, but to find the common ancestor of birds and humans we would have to go back some 250 million years (and it couldn't fly), while the common ancestor of ourselves and the chimpanzees existed a mere 5 or 6 million years ago. I am therefore compelled to adopt the more conventional, down-to-earth view that our descent was not from the creatures of the sky but from the more restricted arboreal heights of our primate forebears. Those seductive parallels between characteristics we fondly imagine to be unique to ourselves and their taunting counterparts in birds are most likely the results of what is known as convergent evolution—independent adaptations to common environmental challenges—rather than features that were handed down from that 250-million-year-old common ancestor. But if there is any one characteristic that distinguishes us from birds, and probably from any other nonhuman creature, it is indeed that extraordinary accomplishment that we call language.

■ The specialness of language ■

Unlike birds, people use language, not just to signal emotional states or territorial claims, but to shape each other's minds. Language is an exquisitely engineered device for describing places, people, other objects, events, and even thoughts and emotions. We use it to give directions, to recount the past and anticipate the future, to tell imaginary stories, to flatter and deceive. We gossip, which is a useful way to convey information about other people. We use language to create vicarious experiences in others. By sharing our experiences, we can make learning more efficient, and often less dangerous. It is better to tell your children not to play in traffic than to let them discover for themselves what can happen if they do.

Even birdsong, for all its complexity, is largely stereotyped, more like human laughter than human discourse. Give or take a few notes, the song of any individual bird is repetitive to the point of monotony. Human talk, by contrast, is possessed of a virtually infinite variety, except perhaps in the case of politicians. The sheer inventiveness of human language is well illustrated in an anecdote involving the behavioral psychologist B. F. Skinner and the eminent philosopher A. N. Whitehead. On an occasion in 1934, Skinner found himself seated at dinner next to Whitehead and proceeded to explain to him the behaviorist approach to psychology. Feeling obliged to offer a challenge, Whitehead uttered the following sentence: "No black scorpion is falling upon this table," and then asked Skinner to explain why he might have said that. It was more than twenty years before Skinner attempted a reply, in an appendix to his 1957 book *Verbal Behavior*. Skinner proposed that Whitehead was unconsciously expressing a fear of behaviorism, likening it to a black scorpion that he would not allow to intrude into his philosophy. (The skeptical reader might be forgiven for concluding that this reply owed more to psychoanalysis than to behaviorism.)

Be that as it may, Whitehead had articulated one of the properties of language that seem to distinguish it from all other forms of communication, its *generativity*. While all other forms of communication among animals seems to be limited to a relatively small number of signals, restricted to limited contexts, there is essentially no limit to the number of ideas, or propositions that we can convey using sentences. We can immediately understand sentences composed of words that we have never heard in combination before, as Whitehead's sentence illustrates.

Here is another example. A few years ago I visited a publishing house in England and was greeted at the door by the manager, whose first words were: "We have a bit of a crisis. *Ribena is trickling down the chandelier*." I had never heard this sentence before but knew at once what it meant, and was soon able to confirm that it was true. For those who don't know, ribena is a red fruit drink that some people inflict on their children, and my first sinister thought was that the substance dripping from the chandelier was blood. It turned out that the room above was a crèche, and one of the children had evidently decided that it would be more fun to pour her drink onto the floor than into her mouth.

This example illustrates that language is not just a matter of learning associations between words. I had never in my life encountered the words *ribena* and *chandelier* in the same sentence, or even in the remotest association with each other, yet I was immediately able to understand a sentence linking them. Rather than depending on previously learned associations, language allows us to connect concepts that are already established in the mind. It operates through the use of rules, known collectively as *grammar.* I hasten to assure the nervous reader that grammar does not refer to the prescriptive rules that some of us struggled with in school but rather to a set of largely unconscious rules that govern all natural forms of human speech, including street slang. In this sense, there ain't no such thing as bad grammar, and it don't really matter what your teacher tried to teach you. Even so, I need to torment you with a short grammar lesson.

■ A grammar lesson ■

Something of the way in which grammar operates to create an endless variety of possibilities is illustrated by a familiar childhood story, in which each sentence is built from the previous one:

> This is the house that Jack built.
> This is the malt that lay in the house that Jack built.
> This is the rat that ate the malt that lay in the house that Jack built.
> This is the cat that killed the rat that ate the malt that lay in the house that Jack built.

. . . and so on, potentially forever, although limited in practice by constraints on short-term memory. In these examples, the phrases qualifying each character in the story are simply added: *the cat that killed the rat, the rat that ate the malt, the malt that lay in the house, the house that Jack built.* But qualifying phrases can also be embedded, like this:

> The malt, which was eaten by the rat that was killed by the cat, lay in the house that Jack built.

And phrases can be embedded in phrases that are themselves embedded, although too much embedding can create a kind of linguistic indigestion that makes a sentence hard to swallow, as in the following:

> The malt that the rat that the cat killed ate lay in the house that Jack built.

This ability to tack clauses onto clauses, or embed clauses within clauses, is known as *recursion*. Mathematically, a recursion formula is a formula for calculating the next term of a sequence from one or more of the preceding terms. Clauses like *that ate the rat* and *that killed the cat* are relative clauses, and a simple formula dictates that a relative clause can be defined (or "rewritten") as a relative clause plus an (optional) relative clause! This formula allows relative clauses to be strung together indefinitely, as in "The House That Jack Built." Grammar is often expressed in terms of rewrite rules, in which phrases are "rewritten" as words and other phrases, and it is this rewriting of phrases as combinations involving phrases that gives grammar its recursive property (see figure 1.1). Perhaps the most minimal example of recursion in literature is that penned by the American writer Gertrude Stein in her poem *Sacred Emily*:

> A rose is a rose is a rose is a rose, is a rose.

This isn't quite as simple, perhaps, as it looks at first glance—note that cunningly placed comma.

It is also clear that *rules* rule, and not just associations. We may learn poems or everyday expressions by heart, simply associating the words together, but when we generate new sentences we do not rely on past associations between words. In the last of the above sentences about the house that Jack built, the words *malt* and *lay* are associated in the meaning of the sentence but are separated by eight other words—and of course even more words could have been inserted, had we chosen, for example, to mention that the rat was fat and the cat lazy. Yet the speaker and the listener both understand that the malt did not kill or eat, but in fact lay in the house that Jack built, at least until greedily devoured by the rat. Our ability to construct and understand sentences depends on a remarkable skill in the use of rules. Even more remarkably, perhaps, we apply these rules without being aware of them, and even linguists are not agreed as to what all the rules are and precisely how they work.

Linguists also like to draw a clear distinction between grammar and meaning. We can understand sentences to be grammatically reg-

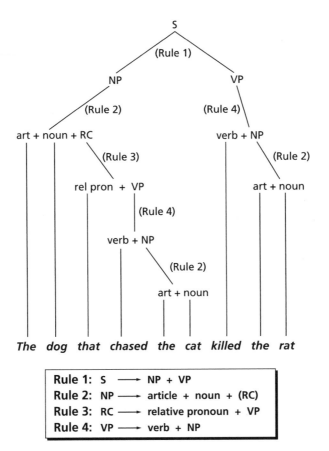

Rule 1: S ⟶ NP + VP
Rule 2: NP ⟶ article + noun + (RC)
Rule 3: RC ⟶ relative pronoun + VP
Rule 4: VP ⟶ verb + NP

■ **Figure 1.1.** ■

The four rules shown at the bottom generate sentences like the one in the upper diagram. Note that the use of rules is recursive. For example, Rule 2 defines a noun phrase in terms of an optional relative clause, defined by Rule 3 in terms of a verb phrase, which is defined by Rule 4 in terms of a noun phrase! This means that you can cycle through Rules 2, 3, and 4 to insert as many relative clauses as you like. *S* = sentence; *NP* = noun phrase; *VP* = verb phrase; *RC* = relative clause.

ular even if they have no meaning, as in the sentence *Colorless green ideas sleep furiously,* constructed by the most eminent linguist of our time, Noam Chomsky. Indeed, we can recognize a sentence as grammatical even if the words don't have any meaning at all, as in Lewis Carroll's "Jabberwocky":

> T'was brillig and the slithy toves
> Did gyre and gimble in the wabe.

> All mimsy were the borogoves
> And the mome raths outgrabe.

But note that some of the words (*was, and, the,* etc.) *are* regular English words. These words are called *function words,* as distinct from the *content words* that refer to objects, actions, or qualities in the world. Suppose we insert nonsense words in the place of the function words:

> G'wib brillig pog dup slithy toves
> Kom gyre pog gimple ak dup wabe.
> Utt mimsy toke dup borogoves
> Pog dup mome raths outgrabe.

Now we have no idea whether this is grammatical or not. This illustrates that function words play a critical role in grammar, providing a kind of scaffold on which to build sentences. Function words include articles (*a, the, this,* etc.), conjunctions (*and, but, while,* etc.), prepositions (*at, to, by,* etc.), pronouns (*I, you, they, it,* etc.), and a few other little things. Content words, by contrast, are easily replaceable, and as speakers we are always receptive to new words that we can easily slot into sentences. We live in a world of rapid invention, and new words, like *geek* and *dramedy* (a drama that doesn't know whether it's funny or not), are coined every day. Another word I encountered recently is *pracademic,* referring to the rare academic who is blessed with practical skills.

Of course, different languages have somewhat different rules, and no one claims that each language has its own set of innate rules. Forming a question in Chinese is not the same as forming a question in English. One important way in which languages differ has to do with the relative importance of word order and what is known as *inflection.* If you studied Latin, you will know that there are many different forms of a noun or a verb, depending on its role in a sentence, and it is these different forms that are known as inflections. In English there are just two forms of the noun, one for the singular and one for the plural (e.g., *table* and *tables*). The Latin word *mensa* means table, but takes several different forms. If it is a direct object (as in *I overturned the table*), it must be rendered as *mensam,* and in the plural (tables) the equivalent forms are *mensae* and *mensas.* Again, the English phrase *of tables* is rendered in Latin as *mensarum.*

The contrast between English and Latin is much more extreme for verbs. There are just four forms of the regular verb in English (e.g., *love, loves, loved, loving*). In Latin there are dozens, as many a struggling schoolchild knows (or once knew). Just to take the present tense, we have

amo	I love
amas	you (singular) love
amat	he/she/it loves
amamus	we love
amatis	you (plural) love
amant	they love

And that's just the beginning of love. There are different forms for the future and past tenses, as well as more complex tenses, like the future perfect (*she will have loved*), the subjunctive, the conditional, and goodness knows what else. In Latin nearly all of this is accomplished by inflecting a basic stem, whereas in English we make much more use of function words (e.g., *they might have loved, she would have been going to love*). In some languages, there are even more variations. For example, Turkish is so highly inflected that there are said to be over two million forms of each verb! The different forms not only reflect the subject of the verb (*I, you, she,* etc.) but also the direct and indirect objects, and a lot else besides.

English is highly dependent on how we order the words. *Man swallows whale* has a rather different meaning from *Whale swallows man*, and is arguably more interesting.[5] But in Latin the subject and object of a sentence are signaled by different inflections, and the words can be reordered without losing the meaning. The Australian aboriginal language Walpiri is a more extreme example of an inflected language in which word order makes essentially no difference; such languages are sometimes called *scrambling languages*. Chinese, by contrast, is an example of an *isolating language*, in which words are not inflected and different meanings are created by adding words or altering word order. English is closer to being an isolating language than a scrambling one.

Given the different ways in which different languages work, it might seem that no set of rules could apply to all of them. Chomsky has neverthe-

[5] *In* The Language Instinct, *Steven Pinker remarks that "Man bites dog" is news, whereas "Dog bites man" is not. He was right. On 18 June, the* New Zealand Herald *carried the headline "Woman Bites Dog," referring to an incident in Tallahassee that evidently made news around the world.*

less argued that certain deeper rules are common to all languages. He refers to these rules as *universal grammar*. One way to conceptualize this is in terms of *principles* and *parameters*. In this view, the universal rules are the principles, and the particular forms they take are parameters that change from one language to another. Although some progress has been made toward identifying universal principles, linguists by no means agree as to what they are, or even as to whether language can be fully understood in this way.

This has not been a complete grammar lesson, but I hope to have illustrated the complexity of grammar and to have demonstrated that grammar operates according to rules rather than simple learned associations. It's true that we do learn some things, like poems, songs, prayers, or clichés, by rote, but this does not explain our extraordinary capacity to generate new sentences to express new thoughts, or to understand sentences—like *Ribena is trickling down the chandelier*—that we have never heard before. It is grammar, then, that gives language its property of *generativity* and distinguishes it from all other forms of animal communication. So far as we know, there is nothing remotely resembling grammar in any of the communication systems of other species: no function words, no recursion, no tenses; indeed, no *sentences*. This is not to say that nothing in the communication or actions of other animals bears on human language, but it is clear that the gap between human and animal communication is very wide indeed, and is one of the greatest challenges confronting psychological science.

■ How is language learned? ■

According to Chomsky, language is too complex to be learned by observation of its regularities. That is, no purely inductive device could possibly extract the rules of language simply by examining or analyzing examples of sentences. Therefore, children must possess some innate knowledge of language that enables them to acquire it, or what Steven Pinker called "the language instinct."[6] In other words, they are born with a knowledge of universal grammar and simply adapt—or "parameterize"—this innate knowledge to conform to the specific language or languages they acquire.

This rather controversial notion at least captures one important

---

[6] *Pinker 1994.*

truth about language: children of any race and culture can learn *any* language, which implies that language does have a universal property. Eskimo children brought up in France will speak French, and visitors to London are often surprised to hear people of African descent speaking English with Cockney accents. Under normal circumstances, all human children learn language, and the languages they learn are the languages they are exposed to in childhood. Of course we *can* learn languages as adults, but only with considerable effort, and it is probably impossible to do so if we have not learned another language in childhood. Another argument in favor of some kind of universal foundation for language is that all languages have the same kind of units, such as nouns, verbs, adjectives, function words, phrases, and sentences. It is also important to understand that the different languages of the world differ very little, if at all, in grammatical complexity. Grammatically speaking, no language is more "primitive" than any other, unless we include languages that are not yet properly formed, like baby talk or pidgin languages improvised by adults to communicate across linguistic boundaries. The grammatical complexity that different languages share is at least consistent with the idea of a common, universal grammar.

But although the acquisition of language is universally human, the fact that languages differ, typically to the point of mutual incomprehension, means that there is of course a learned component. In speech, the actual words we use are arbitrary and must be learned by rote. As we have seen, the rules also vary and depend on experience with the language, although the learning of rules may be more a question of selecting among preexisting alternatives than of rote learning. And although all languages are about equally complex in grammatical terms, they do, of course, vary in terms of the number of *words* they employ. In this respect, English is easily the most bountiful language in the world, in part because it has borrowed vocabulary from a good many other languages, and in part because it has become the main language of science and technology and so must absorb large numbers of new words for different inventions and concepts. This is not to say that English has a monopoly on concepts: some words in other languages express ideas or concepts that have no exact equivalents in English. Not all cultures think alike.

But is it really true that we could not learn language unless it possessed some universal grammatical structure, innately known to us? Chomsky's argument is essentially based on the idea that language is *impossible* to learn from the body of evidence available, and that there *must* be some predetermined structure that guides the discovery of grammatical rules. Consider, for example, how we turn a declarative sentence into a question:

> The brigadier and his wife are coming to dinner tonight.

becomes

> Are the brigadier and his wife coming to dinner tonight?

Here, the rule seems simple: you simply scan the sentence for the word *are* and move it to the beginning. But suppose we apply this rule to a slightly more complex sentence, like

> The brigadier and his wife who are visiting the city are coming to dinner tonight.

This produces the anomalous sentence

> *Are the brigadier and his wife who visiting the city are coming to dinner tonight?[7]

Children virtually never make mistakes like this[8] but seem to understand that it is the second *are*, not the first, that must be moved to create the question:

> Are the brigadier and his wife who are visiting the city coming to dinner tonight?

That is, children seem to instinctively understand the phrase structure of the sentence and so skip over the embedded phrase *who are visiting the city* when making the transformation.

On the face of it, this argument seems compelling, but to conclude that it is *impossible*, without some built-in structure, to learn the rules and phrase structure of grammar may be premature. They once said it was impossible to climb Mount Everest.[9] And it has recently been suggested that language learning may not be so special after all. Since about the mid-1980s researchers have increasingly challenged the idea that the mind is a computational device, operating according to

[7] Linguists have this cute way of putting an asterisk in front of sentences that are anomalous.
[8] See Crain and Nakayama 1986.
[9] Proven wrong, you will be pleased to know, by a New Zealander, along with a Tibetan Sherpa.

rules, and have suggested instead that it is after all merely a sophisticated associative device. It is the brain that creates the mind, and the brain does seem to work by means of elements, called *neurons*, that connect in associative fashion. It is neurons that convey information from the sense organs to the brain, and from the brain to various output devices. The traffic is not all one-way, since there are feedback processes, and circuits in which the firing of neurons is arranged in recurrent loops. Further, we have good evidence that the connections between neurons, known as *synapses*, can be modified by experience, and it is this modification that forms the basis of learning and memory.

Many investigators have tried to create artificial networks that mimic the properties of the human mind, and one of the challenges has been to create networks that demonstrate some of the apparently rule-governed properties of language. For example, Jeff Elman has devised a network with recurrent loops that can apparently learn something resembling grammar. Given a partial sequence of symbols, analogous to a partial sentence, the network can learn to predict events that would follow according to rules of grammar. In a very limited way, then, the network "learns" the rules of grammar. An important aspect of Elman's work is that he makes no attempt to teach the network the rules of language themselves. During training, when the network predicts the next word in a sequence, this is compared to the actual next word, and the network is then modified so as to reduce the discrepancy between them. That is, the network apparently learns to obey the rules without "knowing" what they are: no one programs it to obey them, nor has it been hard-wired to do so.

At first, as one might expect from Chomsky's arguments, the network was not able to handle the recursive aspects of grammar, in which phrases are embedded in other phrases, so that words that go together may be separated by several other words. However this problem was at least partially surmounted when Elman introduced a "growth" factor. Early on, the system was degraded so that only global aspects of the input were processed, but the "noise" in the system was gradually decreased so that it was able to process more and more detail. When this was done, the system was able to pick up some of the recursive quality of grammar, and so begin to approximate the processing of true language. Again, no rules were explicitly taught or built into the system.

Part of the problem in learning grammar has to do with its hierarchical structure. Some of the rules involve the embedding and movement of entire phrases; others, the placement and inflection of individual words; and still others, the component parts of words. The suggestion that arises from Elman's work is that this problem is solved by introducing a growth factor into the network itself, so that it initially processes only global properties of the input but gradually focuses more and more on the details. The developmental psychologist Elissa Newport has characterized this as a "less is more" principle; the reason that children learn language so easily is that they process the information crudely at first, and then gradually acquire detail. Far from being linguistic geniuses, as Steven Pinker has claimed, young children succeed precisely because their learning is diffuse and ill-formed. It is a bit like gradually focusing a telescope; only blurred outlines are visible at first, and then the details gradually emerge.

These ideas, elaborated in the book *Rethinking Innateness* by Elman, Newport, and their colleague Elizabeth Bates, constitute a significant challenge to the idea that humans possess a specific grammar gene, or that language requires some special "language acquisition device."[10] Instead, our unique capacity for language may depend simply on evolutionary alterations to the growth pattern, in which the period of postnatal growth became relatively longer than it is in other primates, the brain grew to a larger size relative to the body, and the relative sizes of different parts of the brain shifted—but more of this later. To be sure, the specific pattern of changes is uniquely human and involves genetic modifications, but they are the sorts of modifications that have altered the basic body plan of animals throughout biological evolution.

If Elman and his colleagues are correct in assuming that grammar can be acquired by means of an associative device that includes a growth component, this does not mean that language does not follow rules. As we saw earlier, language is exquisitely rule-governed, and linguists like Chomsky have done a great deal to show us the nature of the rules. The point is that rule-governed behavior need not require that the rules be pre-programmed into the system, or even represented explicitly in the network. We do not *know* most of the rules that govern our language in any sense other than that we follow them when we speak. The rules

[10] See Elman, Bates, and Newport 1996. Another forthright challenge to the idea that language depends on innate, special-purpose devices is provided by Sampson 1997.

themselves are not associative, but it is possible that they can be learned by an associative device.[11]

That said, it must be recognized that human language is highly complex, and Elman's relatively simple demonstrations do not really come close to capturing many of the niceties of grammar and meaning. Predicting the next word in a sentence is a long way off from actually comprehending a sentence, or producing one. And to be sure, there is something a bit zombielike about a network that responds in a languagelike way but has no built-in rules and no apparent understanding of what it is "saying." But Elman's work does go some distance toward demystifying language and bringing it into the realm of biology, from which it is always in danger of escaping. Clearly it will take a lot more research to convince most linguists that the secret of language learning lies in patterns of growth rather than in special-purpose grammar genes, but that's what the new millennium is for.

Language cannot be entirely dependent on genes, though, because it is heavily influenced by culture. Indeed, we are virtually helpless in a culture where a different language is spoken—unless we resort to gesture, but that's a story for later. One might be almost tempted to believe that language is a mechanism for preserving cultural integrity and keeping foreigners out! Many human characteristics clearly depend not on the genetic code but on the culture we happen to be part of. Richard Dawkins has dubbed these culturally determined characteristics "memes."[12] They include stories, songs, beliefs, inventions, political systems, cuisine—indeed, virtually all of the things we think of as part of culture.

But could language itself be a meme? In some respects, it is. The actual words we use are passed on by the culture we live in, as are accents, catch phrases, and other superficial aspects of language. But language cannot be purely cultural. Whether or not there are "grammar genes," as Pinker maintains, there is no evidence that other species can

[11] The argument between the Chomskyans and the connectionists is also conducted at the level of words themselves. It is clear that the acquisition of vocabulary is largely a matter of rote learning, since words are essentially arbitrary labels that vary from language to language. Yet words also obey rules, as when we form the plural of cat to make cats, or the past tense of stroll to make strolled. A complex word like antidisestablishmentarianism is clearly composed of parts that are familiar from other contexts and ordered according to rules. English is also a highly irregular language, so that the plural of woman is not womans but women, and the past tense of come is not comed but came. In his book Words and Rules (1999), Pinker describes some of the unexpected complexities involved in the formation of words according to rules and argues that the learning of words can be only partially explained in terms of connectionist principles; it also depends on the application of rules that are in part innate. Already, his views have been attacked by connectionists who maintain that Pinker is guilty of viewing words through rule-tinted glasses (McClelland and Seidenberg 2000). For a detailed account of the whole debate, see Clahsen 1999 and the ensuing commentary.

[12] Dawkins 1976. The idea has been expanded in Blackmore 1999.

learn anything resembling true grammatical language, as we shall see in chapter 2. Moreover, memes depend fundamentally on our capacity to *imitate*, which itself is something that humans excel at. As we shall see in the following chapters, even our closest relatives, chimpanzees and bonobos, are relatively poor at imitating. And if that were not enough, true language goes beyond imitation. As I have tried to explain, language is relentlessly generative, allowing us to convey novel thoughts, as I hope I am managing to do in this book.

Another argument for an innate component underlying grammatical language comes from a phenomenon called *creolization*. In the days of colonial expansion, the European traders and colonizers communicated with indigenous peoples with a makeshift form of language called *pidgin*. Pidgin has virtually no grammar—no tenses, no articles like *a* or *the*—but was adequate for the exchange of simple information, as in trading or bartering. Pidgins can become quite complex, but complexity is achieved by stringing words together in associative fashion rather than by the more economical use of syntax. In Solomon Islands pidgin, Prince Charles is known as *pikinini belong Missus Kwin*, and Princess Diana was known as *Meri belong pikinini belong Missus Kwin*, until her divorce, when her title was upgraded to *this fella Meri he Meri belong pikinini belong Missus Kwin him go finish*.[13]

Research in Hawaii has shown that in the course of a generation a pidgin language can be converted into a more sophisticated language, known as a *creole*. Unlike pidgin, a creole does have fully fledged grammar. And it came out of the brains of babes and sucklings, as it were: all it took was for the children of the next generation to be exposed to pidgin at an early age. With no parental help, the children constructed the grammar, presumably because of the intricate grammatical machinery that was already wired into their brains![14]

■ Language, speech, and thought ■

Language is not just speech. We can, of course, read silently and think in silent words. More critically, the signed languages invented by deaf people all over the world have all the generativity of language and are governed by grammar, yet have no basis in sound. They consist entirely of bodily gestures, mostly of the hands, arms, and face. Signed language has all the essential properties of spoken language, includ-

---

[13] *Eerily prophetic.*
[14] *See Bickerton 1984, but see also the critical commentaries that follow.*

You're supposed to
rotate me mentally,
not physically

■ Figure 1.2. ■
Which arm is this agreeable fellow holding out?

ing grammar. I shall examine signed language in much more detail in chapter 6, since it provides one of the foundations for the main theme of this book, which is that even spoken language may have its origins in the silent gestures of our distant forebears.

Language, therefore, runs deeper than speech. Is it the same as thought? It is sometimes suggested that thinking is simply internal speech, and sometimes it is, but not always. There are ways of thinking that owe little to language. For example, we can imagine objects or scenes and manipulate them in our minds. A much studied example is mental rotation, which involves imagining how objects look if rotated into different orientations. Look at this picture of an upside-down man, holding out an arm (figure 1.2). Which arm is he holding out—the left or the right? To answer this question, you may find that you have to mentally rotate the man to the upright position, and perhaps turn him around—processes that have nothing to do with words.

Nonverbal thinking depends on our ability to represent objects, sounds, and actions in our minds and to manipulate them mentally. Besides rotating things, we can replay tunes in our minds or replay a passing shot in tennis or a goal scored in soccer and imagine how we might do these things on some future occasion. Such is the stuff of imagination and fantasy, and words or signs need be no part of it. We use nonverbal thinking to solve problems, and it is likely that our most creative thoughts are nonverbal, and often spatial, rather than linguistic. Albert Einstein, for example, is said to have worked out the theory of relativity by imagining himself traveling on a beam of light. There is no reason to doubt that even the great apes have the capability of forming mental representations of objects and manipulating them. For example, Wolfgang Kohler, in a classic series of experiments, showed that chimpanzees could solve mechanical problems in their minds before demonstrating the solutions in practice, a process he called *insight*.[15]

Language is nevertheless intimately connected with thought, since we use it to convey our thoughts to others. This requires that symbols, whether words or signs, be associated with the objects, actions, qualities, etcetera, that we store in our minds. By manipulating those symbols, we can transmit thoughts from our own minds to the minds of others. This can be effectively accomplished by writing, and I hope these very words are making some sort of impression on your thoughts. Novels and stories are a powerful and compelling means of creating images and fantasies in the minds of others. Television and film, of course, provide direct access to our internal representations, without the need for intervening symbols, except in the case of dialogue.

The language of thought is known as *mentalese*. Not surprisingly, it has much in common with communicative language. Our thoughts are generative, and we can imagine novel scenes, such as a cow jumping over the moon, as readily as we can construct novel sentences to describe them. Our thoughts can also be recursive. For example, one of the characteristics of human thought is what has been called *theory of mind*. This refers to the ability to understand the minds of others and to know what others see, or feel, or know. This can be recursive; for example, I might not only know that you can see me, but I might know that you know that I know that you can see me. The generativity and recursiveness of human language no doubt reflect the generativity and recursiveness of human thought.

But communicative language must be different from mentalese. For one thing, it must make use of symbols to stand for the things we

---

[15] *Köhler 1925.*

want to talk about, since we cannot directly convey our internal representations. The use of symbols requires shared convention; that is, if I am to converse with you I must assume that your understanding of my words is the same as my own. (This in itself implies theory of mind.) Spoken language also differs from mentalese in that it is confined to a single dimension, time. Our thoughts, by contrast, can make use of all four physical dimensions, three dimensions of space and one of time. For example, I can form a three-dimensional spatial image in my mind of the inside of my house from a particular location,[16] but in order to describe it to you I must take a mental walk through the house—a four-dimensional activity—and describe the various features one by one—a one-dimensional activity. This is known as *linearization*, and at least some of the properties of spoken language reflect this requirement. The embedding of phrases may relate to the sort of embedding that can occur as I imagine myself walking through the house; I may stop at a china cabinet, for example, and describe its contents, before continuing to the next item of furniture. That is, the thought processes themselves are hierarchical, ranging from the gross layout of the house, to the items of furniture within the rooms, to the smaller items contained within those items, and so on. One is reminded of Jonathan Swift's comment on fleas:

> So, naturalists observe, a flea
> Hath smaller fleas that on him prey;
> And these have smaller fleas to bite 'em
> And so proceed *ad infinitum*.[17]

Some features of language, therefore, such as its generativity and recursiveness, derive from features of thought itself. The special qualities of speech, at least, derive from the necessity to transform the intended message so that it is transmitted as a signal varying in time. The same sort of transformation occurs in transmitting a TV signal. The spatial pattern is scanned sequentially, so that the pixels on the screen that make up an image are transmitted one at a time and are then recomposed into a spatial pattern at the receiving end. Similarly, we turn our thoughts into a stream of sounds, and the listener then converts these sounds back into the thoughts that we hope to convey. Although we can never

[16] This is not to say that the physical representation in my brain is an exact replica of the house. Rather, the correspondence between physical shapes in the world and their representations in the brain is what has been termed a second-order isomorphism (Shepard 1978).

[17] From On Poetry. Swift was wrong, of course. Fleas aren't molested by smaller fleas.

be quite sure that the listener gets precisely the message we want, the speech system is remarkably powerful, accurate, and flexible.

The linearization problem is not quite so acute in the case of signed language, since the hands and arms can convey something of the spatial quality of the thoughts we might wish to convey, as we shall see in chapter 6. Moreover, where spoken words are arbitrary and depend on convention to convey their meaning, manual signs can in some instances represent shapes and actions more or less directly. The sign for a tree, for example, might depict the actual shape of a tree. Whereas nearly all words represent their meanings symbolically, signs have an imitative or iconic component that may make them easier to learn. It seems reasonable to suppose, then, that there is a more direct relation between signs and the thoughts they express than there is between words and the underlying thoughts. This is but one reason why I shall suggest that language may have originated in manual signs rather than in vocal sounds.

■ Summary ■

In summary, language is an extraordinary accomplishment, and almost certainly a uniquely human one, an idea I hope to amplify in the next chapter. It involves a complex system of rules, and our system of learning those rules is probably innately determined, even if the particular languages we speak have a strong cultural component, to the point of mutual incomprehensibility between cultures. Arguably, it is language that makes us human. Yet such a complex ability cannot have evolved entirely de novo in our species. In the following chapters I shall look closely at the roots of language in our primate ancestry and try to trace how it emerged in the evolution of our own species.

But even at this stage, I hope we can be fairly sure of one thing. Language is not, after all, for the birds.

2 ■ Do Animals Have Language? ■

The seventeenth-century French philosopher René Descartes has been described as the founder of modern philosophy. Some of his ideas arose from a consideration of mechanical toys, which were popular in his day, and he argued that nonhuman animals, even apes, were no more than complicated machines. He also thought that much of the operation of the human body might also be explained in terms of mechanical principles, but not all of human activity, since humans possessed a freedom of will that could not be reduced to mere mechanisms. Language seemed to provide one piece of evidence for this freedom, since there are seemingly no limits to what humans, even human "imbeciles," can say. According to Descartes, the only explanation for this freedom from mechanical constraints was that it must have been bestowed by God.[1]

The idea that the mind cannot be reduced to mechanical bodily processes is known as *mind-body dualism* and is still perhaps the dominant belief in everyday folk psychology. In the 8,958th issue of that venerable British periodical the *Spectator*, regular columnist Frank Johnson rails against boffins who would reduce humans to robots and proudly declares his belief in an immortal soul.[2] Somehow, we feel that we have control over our minds, and our wayward tongues, in a way that goes beyond mere machinery. Descartes's views also engender a comfortable sense that we

[1] Descartes 1647/1985.
[2] Johnson 2000. "Human beings will always top the earthly hierarchy," he declares (11).

are superior to other animals, which no doubt helps appease our guilt for the unspeakable things we do to them. But he did not have it all his own way. He was challenged by a remarkable woman, Princess Elizabeth of Palatine, whose mother was Elizabeth Stuart, daughter of James I of England and sister of Charles I. Elizabeth maintained a friendly correspondence with Descartes in which she questioned his notion that the human mind did not operate according to mechanical laws. Descartes's letters to her were published in 1657, but Elizabeth, a devout but tolerant Calvinist, refused to allow hers to be published, perhaps through fear of offending the church. It was not until over two hundred years later that her letters were found, and they were eventually published in 1879.[3]

Elizabeth, and Descartes too perhaps, may well have had good reason to fear recrimination from the church. In 1747 J. O. de la Mettrie published a book entitled *L'Homme Machine*, in which he argued that all behavior, whether reflexive or intelligent, could be explained in terms of "irritation" of the nerves. As a consequence, he was attacked by the clergy and banished from France and later from Holland, ultimately finding refuge in the court of Frederick the Great of Prussia. And even the animals themselves may have been anxious to show that they, too, were not soulless mechanisms. Here is an extract from a book published in the late eighteenth century: "'I have,' said a lady who was present, 'been for a long time accustomed to consider animals as mere machines, actuated by the unerring hand of providence, to do those things which are necessary for the preservation of themselves and their offspring; but the sight of the Learned Pig, which has lately been shewn in London, has deranged these ideas and I know not what to think.'"[4] Erasmus Darwin, Charles Darwin's grandfather, is said to have believed that pigs would have progressed much further had people not been so fond of bacon.[5]

Perhaps the most direct and lasting challenge to Descartes's dualism was to come, not from the pig, but from Charles Darwin's theory of natural selection. Although Darwin hardly mentioned human evolution in his first book, *Origin of Species*, published in 1859, he was clearly implying that humans shared common ancestors with other species. Again, Darwin's early reluctance to refer to human evolution may have been prompted by his own religious belief, and perhaps by a fear of recrimination. As an under-

[3] *Williams 1978.*
[4] *Sarah Trimmer, Fabulous Histories Designed for the Instruction of Children, 3d. ed. (1788), 71, cited in Thomas 1984, 92.*
[5] *Thomas 1984, 132.*

graduate in Edinburgh, he had heard his friend W. A. Browne present a paper to the Plinian Society in which he propounded a materialistic interpretation of life and mind. The paper was so controversial that the society decided to remove all references to it, including the advance announcement of it in the minutes of the previous meeting.

As Descartes had pointed out, one talent that does seem to distinguish humans from other animals is language, and it was the Oxford philologist Friedrich Max Müller who took up the Cartesian sword. "Language is our Rubicon," he declared, "and no brute will dare to cross it."[6] Darwin replied by suggesting that language must have emerged from the inarticulate cries of animals, which Müller derided as the "bow-wow" theory of language. Given such vituperative exchanges, it is perhaps not surprising that in 1866 the Linguistic Society of Paris banned all discussion of the evolution of language. The Parisian linguists were no doubt also aware that speculation on the basis of flimsy evidence is the stuff of unresolvable controversy and dissent, which they were naturally keen to avoid.

By and large, the ban seems to have worked for almost a century, until Noam Chomsky challenged theorists to address the issue by insisting that language is indeed something possessed only by humans. As we saw in the previous chapter, he echoed Descartes in emphasizing the unique generativity and flexibility of language, commenting that it is "based on an entirely different principle" from all other forms of animal communication.[7] Although an avowed neo-Cartesian,[8] he did not see fit to invoke a higher being to explain language, arguing that it might be understood in terms of computational principles. There is perhaps an irony in this, since the computer is just a mechanical device, albeit a more complex one than Descartes could have envisaged.

In the Chomskyan view, then, human language is somehow special, a sort of eighth wonder of the world. In fact, the biologists John Maynard Smith and Eörs Száthmáry have described the shift from primate calls to human language as the latest of eight major transitions in the evolution of complexity, on a par with the emergence of the genetic code.[9] But the idea that generative language is

[6] *Müller 1880, 403.*

[7] *Chomsky 1966, 78.*

[8] *One of his books is called* Cartesian Linguistics *(Chomsky 1966).*

[9] *Maynard Smith and Szathmáry 1995. The eight major transitions all have to do with the way information is transmitted between generations, and are as follows:*

1. *Replicating molecules to populations of molecules in compartments*
2. *Independent replicators to chromosomes*
3. *RNA as gene and enzyme to DNA plus protein (genetic code)*
4. *Prokaryotes to eukaryotes*
5. *Asexual clones to sexual populations*
6. *Protists to animals, plants and fungi (cell differentiation)*
7. *Solitary individuals to colonies (nonreproductive casts)*
8. *Primate societies to human societies (language).*

unique to humans poses severe problems for any account of how language might have evolved, since its very uniqueness means that no information about its possible antecedents can be derived from studying other species. About 5 or 6 million years ago, the branch of creatures leading to modern humans, the hominins,[10] parted company with the branch leading to modern chimpanzees and bonobos, who are categorized as great apes, along with gorillas and orangutans. If there is no languagelike behavior in these great-ape cousins of ours, then it is highly likely that language evolved only in the hominin family and within the last 5 million years. It is unlikely that chimpanzees have somehow *lost* the capacity for language over that period. Moreover, language leaves very little trace in the fossil remains of our hominin forebears. To be sure, some have claimed to find evidence for language in hominin artifacts, such as tools or body ornamentation, or in migration patterns, or in the size and organization of the brain, but this evidence is at best indirect—although I shall have more to say on these topics in later chapters. If only fossils could talk, or if artifacts included tape recordings![11]

But is language *really* unique to humans? The notion is a controversial one still, and the controversy was never more sharply evident than in the late 1950s. The year 1957 saw the publication of two books on language, one marking the end of an era, the other the beginning of a new one.[12] The behaviorist B. F. Skinner was arguably the most influential psychologist of the time, and his book *Verbal Behavior* was a heroic attempt to reduce language to behavioral principles. It was, in a sense, his swan song (although most of his work was based on pigeons), the culmination of a career spent studying how organisms behave. In this view, language was no more than a complex behavior, ultimately explicable according to the same principles that one might use to explain a pigeon pecking a key for food or a child learning to ride a bicycle.

The new boy on the block was Chomsky. His book *Syntactic Structures* was based on his Ph.D. thesis, and showed that language could not be explained in terms of associations, or of any finite-state device, in which each word can be predicted from the words that precede them. At first, each author was no doubt oblivious of the other, but

---

[10] *Until quite recently, the branch leading to modern humans was known as the* hominid *family, but following Groves 1989 the term* hominid *is now commonly extended to include chimpanzees and gorillas as well. Following contemporary practice, I have therefore adopted the term* hominin *for the branch that diverged from the chimpanzee/bonobo branch.*

[11] *Although if the argument of this book is correct, it would be better to have videos.*

[12] *Both books are more or less unreadable. And probably mostly unread.*

two years later, in 1959, Chomsky published a damning review of Skinner's book that was to change the face of research on language and raise challenges that continue to this day and have yet to be fully met. In subsequent writings, Chomsky has gone on to assert, after Descartes, that true language is uniquely human and quite unlike communication among nonhuman animals.

This was a challenge that was to dominate a good deal of research on primates, and especially on our nearest relatives, the chimpanzee and bonobo, in the latter part of the twentieth century.

■ Animal talk ■

Despite Chomsky's assertions, no one can doubt that animals *vocalize*. The forest and countryside can be a cacophony of sound, even if we disregard P. G. Wodehouse's complaint about "the intolerable screaming of the butterflies." Much of the noise comes from birds and has to do with establishing territorial boundaries, mating, and scaring off predators. Of course, some birds are also wonderful imitators of human speech, and tales of talking parrots are legendary. King Henry VII's parrot is said to have fallen into the river Thames and squawked, "A boat! A boat! Twenty pounds for a boat!" When picked up and taken to the king by a waterman who wanted a reward for his efforts, the bird is said to have advised the king, "Give the knave a groat."[13] But it's not just parrots. As we saw in the previous chapter, many birds have extensive vocal repertoires and can outdo humans in their ability to imitate. But the ability to imitate human speech is of course not enough; we do not grant the power of speech to a tape recorder.

There is at least one parrot, named Alex, who can go beyond imitation. Alex has been taught by Irene Pepperberg to use more than one hundred words to refer to objects and actions and can give simple commands and answer simple questions about the locations, shapes, and even number of objects that are shown to him.[14] This demonstrates an ability to combine words in uncomplicated ways, but Alex's exploits fall well short of anything resembling true grammar. There is no recursion, or tense, or embedding of phrases—or any of the generativity of true language.

In marked contrast to birds, nearly all mammals, with the exception of marine mammals and ourselves, are hopeless at imitating

---

[13] *Cited in Thomas 1984, 128.*
[14] *Pepperberg 1990.*

sounds. Terrence Deacon tells of how he was once walking in front of the Boston Aquarium and was startled to hear a voice yell, "Hey! Hey! Get outta there!" It turned out to be Hoover, the talking seal, now, alas, deceased. Hoover does seem to have been a little unusual, since none of the other seals at the aquarium imitated human speech. Dolphins are excellent imitators and learn very quickly to imitate the whistles of other dolphins;[15] they are also said to be extremely good at imitating human sounds.[16] Dolphins are highly social creatures and apparently use imitation as a way of addressing other dolphins in the group or of recognizing kin. Primates, by contrast, are predominantly visual creatures and have highly specialized perceptual mechanisms for recognizing faces—and this, of course, is why we can immediately recognize a friend at the airport among hundreds of unfamiliar faces.

As for mammals, even primates are poor vocal imitators, although they are certainly noisy enough. Although many of their cries are emotionally induced, they sometimes serve to distinguish between one object and another. For example, vervet monkeys emit separate cries to signal the presence of a snake, a hawk, a leopard, a smaller cat, and a baboon, and when hearing these calls, they act in a manner appropriate to the signaled danger.[17] Close examination of the behavior of these animals when they make or respond to calls suggests that the calls are not simply spontaneous exclamations to release emotion—like a scream of fear or a gasp of surprise. Indeed, it has been argued that the calls meet one of the requirements of language, in that they refer to specific objects; they are *referential*. But this may be true in only a limited way. The calls occur only in the presence of the predators they refer to. We humans, on the other hand, constantly use words in the absence of the things we are talking about and combine them in novel ways to create new meanings.

It may be that a *lack* of voluntary control over warning signals is adaptive, because it makes them hard to fake.[18] Warning signals must be reliable and not subjected to the whim of the occasional animal who might be inclined to "cry wolf." But it is exactly for this reason that primate vocal calls are ill-suited to exaptation for intentional communication. The forelimbs offer much better promise. We share with the other primates a long evolutionary history that shaped the hands and forearms as manipulative devices, specialized for intentional action. Nonhuman primates are tree-dwelling creatures, adapted to swing-

[15] *Janik 2000.*
[16] *Tyack 2000.*
[17] *Cheney and Seyfarth 1990.*
[18] *Knight 1998.*

■ Table 2.1 ■
Different Chimpanzee Vocalizations Identified by Jane Goodall

Roar pant hoot hoot	Arrival pant hoot	Inquiring pant
Spontaneous pant hoot	Bark	Waa-bark
Cough (soft bark)	Hoo	Huu
Food grunt	Food aaa	Copulation scream
Whimper	Squeak	Victim scream
Tantrum scream	SOS scream	Crying
Pant grunt	Pant-bark	Pant-scream
Pant	Soft grunt	Extended grunt
Copulation panting	Nest grunt	Laughter
Wraah		

ing from branches, plucking fruit, catching insects, bringing food to the mouth, and grooming. These actions must be flexible and computed "on-line" to meet the ever-changing demands of the forested environment.

It is difficult to find other examples of primate calls that might be described as referential. Even so, many primate species emit many different kinds of calls. Gelada baboons are said to have at least twenty-two different vocal calls,[19] which have been given labels such as *moan, grunt, yelp, snarl, scream, pant,* and so on,[20] but these exist as wholes and cannot be broken down into interchangeable parts, as human words can, and they are not combined into sequences. Chimpanzees also emit a wide range of different calls. Table 2.1 contains a list compiled by Jane Goodall. A review of the literature suggests that only two chimpanzee calls might be described as referential, although the evidence is equivocal.[21] For example, the arrival pant hoot is said to signal the discovery of food, but some evidence indicates that it may have an altogether more selfish motive. In one study, females never emitted pant hoots, and high-status males did so more often than did low-status males, which suggests that the real reason for the pant hoot was to attract estrous females to the site.[22]

Whether or not this is so, we have little evidence that chimpanzees use vocalizations to signal intentions, or even that their vocalizations are under voluntary control. Goodall records an instance of a chimpanzee who found a cache of bananas and evidently wished to keep them for himself. He was unable to suppress the

[19] Aich, Moos-Heilen, and Zimmerman 1990.
[20] Dunbar and Dunbar 1975.
[21] Hauser 1996. He was able to find no examples of referential calls for the other great apes, and even the chimpanzee food calls must be considered only marginally referential.
[22] Clark and Wrangham 1993.

excited pant hoot signaling the discovery of food, but attempted as best he could to muffle it by placing his hand over his mouth. Conversely, it may often be equally difficult for chimpanzees to produce a call on demand. After many years of observing chimpanzees in the wild at Gombe, Goodall concluded: "The production of sound in the *absence* of the appropriate emotional state seems to be an almost impossible task for a chimpanzee."[23]

Chimpanzees do emit sequences of calls that can be quite long and consist of several call types, and extended exchanges often takes place between individuals who are out of visual contact with one another. One might be tempted to think that this is some kind of dialogue. Detailed analyses of the sequences of calls during vocal exchanges show, however, that they have none of the properties of conversation. When people converse, they tend to choose words different from those they have just heard—the response to a question is not the same as the question itself. Even from an acoustic point of view, human conversation consists of the alternation of sounds that are in general dramatically different from one another, whereas chimpanzees produce sequences that tend to be similar to what they have heard. These exchanges probably have to do simply with maintaining contact. Many other primates, including baboons and gorillas, also exchange calls that are acoustically similar, sometimes producing chains of vocalizations across the forest. Calls are sometimes synchronized in choruses. If anything, these phenomena resemble singing, but have little in common with human conversational language.[24]

Joseph Addison, the seventeenth-century English essayist, wrote: "If we may believe our logicians, man is distinguished from all other creatures by the faculty of laughter"[25]—which simply shows that you can't always rely on logic. Laughter is actually common to many species, including (some) humans. Tickling causes even rats to laugh, one recent study claims.[26] In primates, at least, laughter clearly has a social function, since one source of laughter is tickling, and you can't tickle yourself. Actually this is not quite true. You can tickle yourself with a specially designed tickle machine, which introduces a delay between inducing the tickle and having it delivered to the body. With zero delay, there is evidently an internal process of cancellation that neutralizes the tickle, which is why we can scratch

[23] Goodall 1986, 125.
[24] See Arcadi 2000 for a detailed study of how chimpanzees respond vocally to one another and for a review of data on other primates.
[25] Spectator, no. 494, 26 September 1712.
[26] Panksepp and Burgdorf 1999.

under our arms without laughing to death. But with increasing delay the sensation of being tickled increases, reaching a maximum at about one-fifth of a second.[27] That mention of laughing to death was no joke: Simon de Montfort, the thirteenth-century English earl, executed captives by tickling the soles of their feet with a feather. The sustained, uncontrollable laughter eventually caused death through cardiac arrest or cerebral hemorrhage. Laughter is of course involuntary, and it takes a very good actor to produce simulated laughter that sounds like the real thing. Like the pant hoot of the chimpanzee observed by Jane Goodall, laughter can be impossible to suppress: in 1992 a girls' boarding school in Tanzania had to be closed because of an uncontrollable epidemic of hysterical laughter.[28]

But it is of course speech, not laughter, that causes humans to stand out—although one might say that speech creates many more opportunities for comic relief. In evolutionary terms, we are closest to chimpanzees and bonobos, but their vocalizations, whether pant-hooting or simply hooting with laughter, are no more complex than those of other primates.[29]

Although primate vocalizations are largely fixed and tied to specific situations or emotional states, this does not mean that they cannot be modified. Some studies suggest that chimpanzees' food calls are variable, which would indicate a degree of flexibility,[30] although Michael Tomasello has remarked that the observed variation is probably not under voluntary control and may reflect differences in emotional arousal rather than the effects of learning.[31] The long-distance pant hoot of the chimpanzee shows different acoustic patterns in different regions of Africa, in much the same way that birdsong shows dialectic variations, which suggests a cultural influence.[32] But more telling, perhaps, are the regional variations between pant hoots in different captive colonies in the United States, even though the animals in each colony come from different parts of Africa. That these colonies develop their own characteristic dialects, rather than preserving those of their native terrain, seems to indicate that the acoustic patterns are learned.[33] The researchers who documented this argue, though, that the mod-

[27] *Blakemore, Wolpert, and Frith 1998.*
[28] *See Robert Provine's excellent, and indeed humorous, book* Laughter: A Scientific Investigation *(2000).*
[29] *Mitani 1996.*
[30] *Hauser and Wrangham 1987; Hauser, Teixidor, Field, and Flaherty 1993.*
[31] *Tomasello and Call 1997.*
[32] *Arcadi 1996.*
[33] *Marshall, Wrangham, and Arcadi (1999) studied pant-hoot vocalizations in two captive facilities, Lion Country Safari and North Carolina Zoological Park. Not only did the pant hoots differ systematically between these facilities, but an acoustically novel pant hoot that was introduced by a male chimpanzee to the Lion Country Safari spread to five other males in the same colony.*

ified calls are shaped not by imitation but rather by what they call "action-based learning." The pant hoots of young chimpanzees tend to be quite variable, and distinctive calls can be shaped by selective social reinforcement. But this mechanism would be dreadfully slow and inefficient compared to the manner in which human children learn words. It has been estimated, for example, that a child's vocabulary increases dramatically from just a few words to anywhere from ten to fifteen thousand words between the ages of eighteen months and five years. That corresponds to a rate of about one new word every waking hour.[34] This acquisition is largely independent of social reinforcement; it probably involves the remarkable power of humans to imitate,[35] and perhaps an enhanced plasticity of the brain during a period of rapid growth. My own static brain certainly can't pick up words at anywhere near that rate.

Moreover, the main modifications that occur in the pant-hoot cries of chimpanzees are not so much in the actual sounds as in their temporal structure, or timing. Pant hoots are also frequently accompanied by drumming, in which the animals repeatedly hit their hands and/or feet against a variety of surfaces, including their chests, the ground, tree trunks, and the buttresses of trees. Drumming on buttresses produces the loudest sound, louder than the pant hoots themselves, and seems to be a way of maintaining contact with other individuals in the group. Groups have characteristic temporal patterns in their drumming, as do individuals, and these differences may serve effectively as identification tags.[36] Chest beating is also well documented in mountain gorillas, often as an aggressive display, accompanied by threatening motions and vocalizations, but in some cases as a response to chest beating from another unseen individual.[37] The synchronization and association of manual drumming with repetitive vocalization may provide a clue to the link between gesture and vocal language in hominin evolution, although it probably has more to do with group identity than with language. The human equivalent, in modern times, might be a rock concert.

■ Teaching language to animals ■

Animals can't converse naturally in the wild, but we humans have long nursed a fantasy that they might might acquire speech. Chil-

[34] Pinker 1994.
[35] Tomasello 2000.
[36] Arcadi, Robert, and Boesch 1998.
[37] Schaller 1963.

dren's literature, in particular, is heavily populated with talking bears, rabbits, and other cuddly creatures. But the thought that animals might speak to us is not always a comfortable one. In Saki's short story "Tobermory," the weekend guests at a country house were disconcerted when the house cat began to speak, revealing some of the goings-on between the guests and remarking on the hostess in her absence. All were relieved when Tobermory was killed in a skirmish with the big tom from the rectory. Nevertheless, Tobermory may well take the prize for the most articulate talking animal in fiction. Here's what he said when one of the women asked for his opinions of her intelligence:

> "You put me in an embarrassing position," said Tobermory, whose tone and attitude certainly did not suggest a shred of embarrassment. "When your inclusion in this house-party was suggested Sir Wilfred protested that you were the most brainless woman of his acquaintance, and that there was a wide distinction between hospitality and the care of the feeble-minded. Lady Blemley replied that your lack of brain-power was the precise quality which had earned you your invitation, as you were the only person she could think of who might be idiotic enough to buy their old car."[38]

Fortunately for us all, there is probably little prospect of real cats learning to speak, but primates might be a different proposition. Although primate vocalizations in the wild do not bear any real resemblance to human language, it need not follow that one cannot *teach* language to an ape. The idea is actually an old one. In 1661 Samuel Pepys saw a strange creature, probably a chimpanzee or gorilla, in Guinea, and wrote that it was "so much like a man in most things . . . I cannot believe that it is a monster got of a man and a she-baboon. I do believe it already understands much English; and I am of the mind it might be taught to speak and make signs." As we shall see below, these were prophetic words.

Over the past half-century, spurred in part by the Chomskyan challenge, there have been a number of well-publicized attempts to teach language to chimpanzees and other great apes, with at least some success. It quickly proved futile, however, to try to teach apes to actually *talk*. Cathy and Keith Hayes, a wife-and-husband team,

---

[38] *Saki 1930, 130.*

begins at the start

raised a chimpanzee called Viki in their own home from the age of three days to about six and a half years, treating her essentially as one of their own children. Viki was never able to say more than about three or four crudely articulated words: *mama, papa, cup,* and possibly *up.* But Viki's inability to speak need not imply an inability to learn *language.* Another husband-and-wife team, Allen and Beatrice Gardner, noticed from films of Viki that she seemed reasonably intelligible without the sound track. She was often able to get her mouth into more or less the right position to articulate words but could not actually produce the corresponding sounds. What Viki was probably doing was trying to communicate using *visual* imitation.

Spurred by these observations, the Gardners hit upon the idea of trying to communicate with chimps using manual gestures, loosely based on American Sign Language (ASL).[39] They were able to teach well over 100 gestures to another young chimpanzee, named Washoe. Later, Francine Patterson claimed to have taught at least 375 different signs to a gorilla named Koko,[40] enough, she claimed, to enable her to test his IQ, which turned out to be about 90. Another accomplished signer is an orangutan.[41] These studies give immediate and powerful support to the major thesis of this book, which is that human language evolved first as a system of manual gestures. The chimpanzee, along with the bonobo, is our closest relative among the great apes. This makes it likely that the common ancestor of ourselves and these two species, dating from about 5 million years ago, would have been much better equipped to develop a communication system based on manual and bodily gestures than one based on vocalization.

In his book *The Language Instinct,* Steven Pinker recounts how Jane Goodall once remarked that many of the signs used by these apes were familiar to her from her observations of chimpanzees in the wild. Pinker takes this as indicative of the futility of trying to teach signed language to apes, but it might equally be taken as evidence that language in fact evolved from manual gestures, and that the roots of language might indeed be found in our primate forebears.

More recently, though, investigators have replaced gestural communication with systems of arbitrary shapes to denote objects and actions. One reason for this is that there is no ambiguity in the symbols, and they are easy for the animal to manipulate. It is easy for the experimenters to determine which symbols the animals choose,

[39] *Gardner and Gardner 1969.*
[40] *Patterson 1978.*
[41] *Miles 1990.*

whereas gestures are often difficult to decipher, at least for humans. The most impressive results have come from a young bonobo called Kanzi, studied by Sue Savage-Rumbaugh of the Yerkes Laboratory.

Kanzi has shown a remarkable ability to use symbols on a specially designed board to generate messages, and to interpret messages generated by others. The symbols, known as *lexigrams*, are specially chosen *not* to be pictorial representations of what they stand for. That is, they are abstract symbols, containing no clue in the symbols themselves as to what they might mean. Their meanings must therefore be learned by rote, just as human children must learn the meanings of spoken words. As Kanzi learned new symbols, so the keyboard grew, and when the number of symbols reached 256, his minders decided not to add further to it, as it would then become too unwieldy. Kanzi has spontaneously learned to supplement the symbols with manual gestures to expand his vocabulary. He also uses a few vocalizations, but these seem to be emotional rather than denotative. One example is a kind of whine, of the sort that small children use when they badly want something.[42]

Kanzi is able to generate novel requests by pointing to appropriate combinations of symbols on the board, and also to understand novel sequences. These are simple, consisting of two- or three-word combinations like *hide peanut, chase you, hot water there*, or *childside food surprise*. It is reasonably clear from accounts of these exploits that Kanzi is not prompted in any way, and that many of his "utterances" are novel combinations that neither he nor his trainers have used before. Sue Savage-Rumbaugh has claimed that his two-word utterances display a kind of grammar, since the order in which he places the words follows simple rules. For example, he sometimes uses three-word sentences in which there is an agent, an action, and a recipient, as in *you chase Mulika*, where he follows English word order to indicate who chases whom. In two-word combinations involving an action and an object, Kanzi typically puts the action first even in cases where the English order is reversed, as in *chase you* instead of *you chase*, but he indicates the agent by pointing. That is, if Kanzi signs, "Chase you" and points at you, then it is you who must do the chasing.

As is the case with human children, Kanzi's comprehension goes beyond his ability to produce utterances, and he has even developed a

[42] *Kanzi's exploits are described in some detail in Savage-Rumbaugh, Shanker, and Taylor 1998.*

striking ability to understand spoken English. This is tested by giving him spoken instructions, often involving ten or more words, and recording his ability to carry them out. For example, when asked, "Would you put some grapes in the swimming pool?" he immediately complied, getting out of the water, fetching some grapes, and tossing them into the water. On another occasion he was visiting Austin, a chimpanzee, and was told, "You can have some cereal if you give Austin your monster mask to play with." He immediately found his mask, gave it to Austin, then pointed to Austin's cereal. Of course, he is not always correct, but Savage-Rumbaugh describes an experiment in which he was given a list of 660 unusual spoken commands, some of them eight words long, and Kanzi was able to carry out 72 percent of them correctly. Kanzi was nine at the time and did a little better than a two-and-a-half-year-old girl called Alia, who managed 66 percent.

It would be easy to overestimate Kanzi's linguistic skill from these examples, since Kanzi need not process every word and almost certainly did not do so—nor did Alia. The sentences can generally be understood if one extracts only the content words, ignoring the function words. For example, a sentence like *Go get the balloon that's in the microwave* can effectively be reduced to *Get balloon microwave*, and the meaning inferred with little ambiguity. It is easier and more natural to put a balloon into a microwave than a microwave into a balloon, don't you think? Even so, Kanzi's ability to pick out the content words from the more or less continuous sequence of sounds is impressive and unexpected.

Other chimpanzees and bonobos, including Kanzi's mother, Matata, have failed to match Kanzi's achievements, and in fact Matata was so poor at even learning the symbols on the keyboard that her minders were tempted to abandon the whole project. Even more remarkably, Kanzi was not explicitly taught; he spontaneously picked up his skills while watching others using the keyboard and listening to the humans speak. It is unlikely, though, that Kanzi is an exceptional genius. The secret to his success almost certainly lies in the fact that he was exposed to these linguistic items from the age of six months, when he arrived at the research center with his mother. Just as human children spontaneously learn language if exposed to it from a very young age, so Kanzi appears to have been exposed precisely when his growing brain was most receptive to this kind of experience.

Given Kanzi's skills, Sue Savage-Rumbaugh claims to have reasonable conversations with Kanzi, in which he demonstrates memory of past events, such as where he has left a ball; future intentions, like a route he intends to take through the woods to get to some place; and even "theory of mind," an awareness of the feelings of others. For all intents and purposes, Kanzi seems the equivalent, in both linguistic and social skills, of a two-and-a-half- to three-year-old human child—except, of course, that he cannot *talk*.

But does Kanzi have *language*? Steven Pinker, in *The Language Instinct*, is scornful, declaring that even Kanzi just doesn't "get it."[43] And in a sense he is right. Kanzi's productions exhibit very few of the hallmarks of grammar, save for a weak adherence to some simple rules about word order. He makes no use of function words, inflections, or tenses, has not mastered the pluperfect or the future conditional. He does not seem to distinguish between statements, questions, and commands. So far as we know, Kanzi cannot tell a story, although he does appear able to tell a lie—which I suppose is more than George Washington could do (assuming he was telling the truth).

We have another reason to be skeptical of claims about Kanzi. In 1904 a retired schoolteacher named Wilhelm von Osten claimed that a horse, known as "Clever Hans," was capable of humanlike thought and language. He had taught Clever Hans to answer questions by tapping out letters of the alphabet with a front hoof, with each letter represented by a different number of taps. In this manner, the animal was apparently able to answer quite sophisticated questions. For example, when asked "What is 2/5 plus 1/2?" Hans stamped his foot nine times, paused, and stamped his foot another ten times, apparently indicating that the answer was 9/10. Many were convinced, including the leading German psychologist of the day, Professor Stumpf of the University of Berlin. Eventually, one of Stumpf's research students, Oskar Pfungst, demonstrated that von Osten was signaling the animal when to stop tapping by making a slight upward jerk of his head. Even though he was unaware of it, von Osten was himself giving the answers. Nevertheless, this celebrated case convinced a number of prominent scholars of the time that animals could think and even use language, if only instructed in the right way.

Is clever Kanzi another Clever Hans? It is true that those working with him have developed a close relationship, and in some situations

---

[43] *Pinker 1994, 340.*

he may be responding to subtle nonverbal signals, but the scientists studying him are aware of this possibility and have set up reasonably objective tests. I suspect, in fact, that Kanzi has not been given quite the respect he deserves. There remains a strong Cartesian impulse to deny human characteristics to nonhuman animals, and the embarrassment caused by Clever Hans may well have led scientists to overreact to claims of humanlike intelligence in animals. Further, Noam Chomsky's strong position on the uniqueness of human language has dominated linguistics for almost half a century. Now that we're in a new century, we may be able to yield a little, and give animals their due.

■ Protolanguage ■

One way to describe Kanzi's ability to communicate is that it constitutes what the linguist Derek Bickerton has called *protolanguage* rather than true language.[44] Protolanguage has at best a primitive syntax, allowing different combinations of words representing objects and actions. In Kanzi's case, "sentences" may be two or three words long, typically involving a command, an object, and a location, as in *Put (the) grapes (in the) swimming pool.* Bickerton suggests that protolanguage is not unique to Kanzi, or even to the bonobo. As we have seen, other great apes, including several chimpanzees, a gorilla, and an orangutan, have been taught to use gestures and appear able to combine them. It may be reasonable to infer that at least the potential for protolanguage was present in the common ancestor of the great apes (including, of course, ourselves) who was lumbering around Africa some 16 million years ago.

We may also see something like protolanguage in two species of marine mammals, namely, dolphins and sea lions, as well as in Alex, the African gray parrot introduced earlier. This suggests that protolinguistic ability may have evolved independently in at least three evolutionary clades: birds, marine mammals, and great apes. This does not mean that any of these animals use protolanguage in the wild, since the only known examples to date of "talking" animals were taught by humans or, like Kanzi, learned by observing other animals being taught by humans. Protolinguistic ability may, therefore, depend on a more general cognitive ability that enables these animals to form representations in their minds and combine them in meaningful ways. It

---

[44] *Bickerton 1995.*

may be restricted to animals that have adapted to environments in which there is at least a moderate abundance of objects to manipulate.

Given the emergence of protolinguistic ability in our own evolutionary forebears, it is likely that it was the precursor to the languages we speak and sign today. Indeed, we sometimes resort to protolanguage, as when we send (or used to send, before email) telegrams or write headlines. Something like protolanguage can occur as a result of brain damage, in a condition known as *agrammatism*. Other examples include the pidgins that early traders used to speak with the "natives," and my feeble efforts to communicate in Italian. Protolanguage is the language of the two-year-old child, and of the drunken teenager. But it has none of the characteristic scaffolding that we see in fully formed sentences. Protolanguage is not *language* in the sense that I tried to convey in the previous chapter. Even so, we should perhaps not underestimate its power. It is generative, allowing the production and comprehension of novel utterances—although it does not provide anything like the flexibility and *narrativity* of full-fledged syntax.

Why does the linguistic prowess of captive apes like Kanzi seem so much more sophisticated than that of apes observed in the wild? Part of the reason may be simply that the natural communications of apes have not been decoded. As we shall see in chapter 3, a good deal of that communication is probably gestural, involving subtle signals of the hands, face, and body. Remember that Clever Hans, the horse, picked up minute visual cues that clever humans were unable to detect, and that even his trainer, Wilhelm von Osten, was unaware of giving. Another reason, though, may be that animals in the wild simply do not have much to communicate about. Most animal communications seem to consist either of single signals or of random variations on a theme, as in the case of birdsong. If we suppose that animals communicate to signal events to one another, such as the dangerous presence of a snake, then a single signal will generally be sufficient, and will have the advantage of economy.

In a simple world each such event can thus be represented by a single call, but in a complex world it might be more efficient to learn separate symbols for different components of an event. The simplest events consist of objects and actions, such as *baby screams, snake approaches*, or *apple falls*. Suppose, for example, that an animal's experience includes five meaningful objects and five meaningful actions. If each object is associated with a single action, so that only babies scream or only apples fall, then there are only five events to be signaled, and five "event" symbols will do the trick; the objects do not need to be distinguished from the actions associated with them. But if all possible combinations of objects and actions can occur, then it would be more economical to learn five symbols for the objects and five for the actions, making ten in all, than to learn twenty-five symbols to cover all their possible combinations. This might be the source of protolanguage, leading eventually to grammar.

Even a simple system of this kind comes at a price. First, it requires an extension of short-term memory, so that it can process pairs of symbols, not just single, holistic symbols. If one signals "baby bathes," the listener must remember the *baby* while processing the *bathes*, in order not to throw out the baby with the bathwater. Second, one must code not only the symbol but also the category it belongs to, namely, object or action. But at some point in an increasingly complex world of objects and actions, the benefits outweigh the costs, simply in terms of the economy of representation. In a world of ten objects and ten actions, one need only learn twenty symbols to describe the one hundred possible combinations.

A combinatorial system has another advantage. Even in a world of ten objects and ten actions, it is unlikely that all possible combinations will actually be encountered. Apples do not scream, and snakes do not fly. Hence there may be only, say, forty combinations that would normally require symbolic representation. But by learning the names of the individual objects and actions, one can then describe novel and unlikely events: one can recount to an amazed audience how a snake screamed or a cow jumped over the moon. With a primitive grammar, consisting only of symbols for objects and actions, combinable into two-word sentences, generative language is born.[45] This analysis is of course oversimplified, but the same principles can be extended to more complex worlds, where events to be communicated include several objects and actions, different places and times, and so forth.

[45] These ideas are discussed in more detail in Nowak, Plotkin, and Jansen 2000. They compare the costs of learning words for objects and actions and learning words for holistic events, and also consider other variables, such as how many combinations are encountered in an animal's experience and the population dynamics of the spread of words through a linguistic community. In one scenario, in which it is twice as hard to learn a combinable signal as to learn a holistic one, a third of all noun-verb (object-action) combinations describe meaningful events, and there are equal numbers of objects and actions, it follows that a system of eighteen objects and eighteen actions is required for grammar to have a chance of evolving.

In the natural world of the bonobo, there may simply have been too few objects and actions worth talking about for grammar to emerge. Yet the fact that Kanzi and other great apes have been successfully taught protolanguage suggests that they already possessed the *capacity* for it. This capacity probably derives from a capacity for combinatorial *thought*. Kohler's experiments on problem solving in chimpanzees, mentioned in the previous chapter, suggested that the animals were capable of forming representations of objects and actions and combining them in imagination to solve problems. The selective pressure to combine internal representations in this way probably had more to do with solving practical problems than with communicating about them.

Michael Tomasello describes a recent study in which chimpanzees were shown an object that was out of reach and then given a rake. They were able to "solve" the problem and use the rake to obtain the object. However, the chimpanzees did not learn by imitating others. There were actually two ways to do it; one group of chimps observed the first way, and another group the second. But the different ways the chimps actually raked in the objects bore no systematic relation to what they had observed. That is, the information was not communicated even by imitation; the chimps preferred to work it out their own way. When human children were given the same task, they were much more likely to copy exactly what they had seen. This work might be taken to imply that the chimpanzee has the mental capacity to combine internal representations, such as an object and the act of raking it, but seemingly no ability, or inclination, to imitate how others do things.[46]

It must also be said that although chimpanzees can learn to do mechanical tasks, and apparently demonstrate "insight," they perform quite poorly compared with human children. One can be just as impressed by their often apparently obtuse refusal to solve a problem as by their occasional successes.[47] Daniel Povinelli taught chimpanzees to use hooked tools to reach through holes in a plexiglas screen to retrieve a banana that was just out of reach. The banana was on a piece of wood with a vertical post at one end and a ring at the other, and the chimpanzees were rewarded if they hooked the ring and hauled the banana into reach. But when the ring was removed, the chim-

[46]See Tomasello 1996.

[47]We might remember Samuel Johnson's point about a dog walking on its hind legs: it may not do it well, but the wonder is that it can do it at all. Johnson's observation was actually made in another context that might be considered sexist, so I won't repeat it here.

panzees did not seem to understand that they could use the tool to hook the post, and pull the banana in that way. According to Povinelli, chimpanzees actually have rather little understanding of the physical world.[48] Perhaps it is a limitation of this sort that prevents them from going beyond protolanguage to grammar.

■ Some conclusions ■

After some fifty years of trying to teach language to apes, the experts are still divided as to what the results add up to, perhaps surprisingly so. Nevertheless, I think two important general conclusions can be drawn. First, there is little point in trying to get apes to talk. Kanzi is surprisingly good at understanding human speech but apparently cannot produce it. His shrill cries bear no obvious resemblance to spoken words and are probably emotional accompaniments to his attempts to communicate by gesturing and selecting lexigrams.

Second, the great apes, at least, can communicate pretty well through visual means. They can both use and interpret gestures, including facial expressions, and they can communicate with artificial symbols that they and their interlocutors manipulate or simply point to. There is little doubt that communicating visually in these ways is intentional, and not simply dependent on emotional state. Indeed, Jane Goodall's example of the chimp trying to suppress the food call nicely illustrates the distinction between the involuntary call and the deliberate *gesture* of trying to suppress it with the hand. Chimpanzees cannot lie through their teeth, as it were, but they can pull the wool over our eyes.

Through visual means, it is possible not only to establish ways of representing actions, objects, and locations but also to build a protolanguage that uses combinations of symbols to represent commands, or events, or even wishes, in creative ways. To be sure, such a system does not have the flexibility of a fully grammatical language, but it's a start. Moreover, despite the din of the forest, evidence is mounting that the closest approximation to human language among apes in the wild is to be found in their gestures, and not in the sounds that they make. But that is a matter for the next chapter.

---

[48]*Povinelli 2001.*

In the study of humanity there is a maine deficiencie, one
Province not to have been visited, and that is Gesture.
—John Bulwer, *Chirologie* (1644)

3 ■ In the Beginning Was the Gesture ■

We generally take it for granted that the essence of language is
speech. To be sure, we can read and write silently, but written language is parasitic upon speech. Children do not learn to read or write
until they have learned to talk. And while speaking comes naturally
and effortlessly to every normal child, learning to read is often a
painful process, and some otherwise normal people simply never
master it. We evolved to speak, you might think, but reading is a burden imposed upon us by culture. So, take another deep breath,
friend, and read on.

How on earth could speech have evolved? One cannot but wonder how it was invented, since it consists of arbitrary sounds that bear
no relation to the objects, actions, or qualities they represent; for
example, the word *dog*, as spoken, bears no resemblance to that
friendly animal, or to the sounds that it makes. Of course, there are
a few exceptions—onomatopoeic words like *buzz, hum, shriek,* or
zanzara (the Italian word for mosquito, a creature often heard but
not seen)—but for the most part the actual sound of a word gives no
clue as to its meaning. It has been argued that the
earliest words did in fact mimic their referents, a
theory incorporated into what Max Müller
scathingly referred to as the "bow-wow" theory.[1]

[1] *Müller 1880. Müller does give
some consideration, however, to
the view of some philosophers,
such as Adam Smith and Dugal
Stewart, who argue—in the
spirit of this book—that
communication originated in
"gestures of the body, and in
changes of countenance" (33).*

However, this is regarded as fundamentally implausible, since the vast majority of objects, actions, and qualities that we talk about are not associated with sounds of any description. Like Victorian children, and unlike the mosquito, they are seen and not heard. Moreover, words referring to the same objects differ from language to language; a *dog* in English is a *chien* in French, *Hund* in German, *kuri* in Maori.

The mystery of how such a system could ever have *begun* is well captured by Jean-Jacques Rousseau's famous paradox: "Words would seem to have been necessary to establish the use of words."[2] Having pooh-poohed the bow-wow theory,[3] Müller complained that "no one has yet explained how, without language, a discussion, however imperfect, on the merits of each word, such as must needs have preceded a mutual agreement, could have been carried on."[4] Philip Lieberman argued in his book *Eve Spoke* that true spoken language emerged with modern humans, also known as *Homo sapiens*, some 150,000 years ago. The evidence for the emergence of the species *Homo sapiens* itself is based in part on present-day variation in mitochondrial DNA (mtDNA), a form of DNA that is passed down the female line. By extrapolating back in time, and taking into account the rate at which mtDNA mutates, scientists have inferred that all present-day mtDNA is descended from a woman who probably lived (and loved) in Africa about 150,000 years ago.[5] This woman has become known, not surprisingly, as Eve, but if it was she who first spoke, we must surely wonder who could have understood her.[6]

Consider what a useless thing a dictionary is to one who knows nothing of a language. I'm not just referring to Ambrose Bierce's definition of a dictionary as "A malevolent literary device for cramping the growth of a language and making it hard and inelastic."[7] The situation is worse than that. Every word in

[2] "La parole paroît avoir été fort nécessaire pour établir l'usage de la parole" (Rousseau 1775/1964, 148–49).

[3] Müller also pooh-poohed the pooh-pooh theory, which holds that language derived from involuntary interjections.

[4] Müller 1880, 34.

[5] The most recent estimate, which, unlike earlier analyses, is based on the entire mitochondrial genome, is 171,500 plus or minus 50,000 years ago (Ingman, Kaessmann, Pääbo, and Gyllensten 2000). This work offers broad support for the "out-of-Africa theory" proposed by Stringer and Andrews 1988, but there continue to be trickles of doubt (see, for example, Wolpoff, Hawks, Frayer, and Hunley 2001). Mitochondrial DNA taken from an ancient Australian fossil, anatomically modern in form, is apparently more ancient than that of any living person, including contemporary Africans (Adcock, Dennis, Easteal, Huttley, Jermiin, Peacock, and Thorne 2001). Evidence from the Y chromosome, which males jealously keep as their own, also suggests a more recent date for the origins of our species (Underhill, Shen, Lin, Jin, Passarino, Yang, Kauffman, Bonne-Tamir, Bertranpetit, Francalacci, Ibrahim, Jenkins, Kidd, Mehdi, Seielstad, Wells, Piazza, Davis, Feldman, Cavalli-Sforza, and Oefner 2000)—and who could quarrel with such a phalanx of authors? These matters will be discussed further in chapters 5 and 7.

[6] This analysis is a little misleading, since there were undoubtedly other women around in Eve's time, but their mitochondrial genotypes have disappeared. We have no reason to think that Eve herself was special. Even so, it is difficult to understand how language could have emerged de novo in this society—which would also, incidentally, have included men. The chaps did not pass on any mtDNA, but they certainly helped the women do it.

the dictionary is defined in terms of other words, so a dictionary is nothing more than a giant tautology. To get the thing off the ground, there must be some way of indicating what words refer to in the real world. Even Samuel Johnson, pioneer of the modern English dictionary, seems to have understood the problem: "I am not yet so lost in lexicography," he said, "as to forget that words are the daughters of earth, and things are the sons of heaven." Yet, in his perverse fashion, he seems to have it round the wrong way, since things are of the earth, while words have an evanescent, arbitrary quality that might well have been crafted in heaven. But in either case, how were links formed between those arbitrary sounds we call words and the stuff of the real world—a real world made available to us largely through vision and touch, rather than through sound? It seems almost inevitable that those links involved *gesture*. In his novel *One Hundred Years of Solitude*, Gabriel García Márquez writes: "The world was so recent that many things lacked names, and in order to indicate them it was necessary to point."[8]

I began this chapter by pointing out (*sic*) that we generally take it for granted that the essence of language is speech. This assumption is not quite right, since there is a silent form of language that is just as natural to those who learn it. It is called signed language, and is carried on with gestures of the arms, upper body, and face. Deaf communities all over the world have invented signed languages, often in the face of condemnation from deaf educators. The growing recognition of signed languages as true languages, with all of the expressiveness and generativity of spoken language, has provided a powerful boost to the idea that language originated as a gestural system, and may even have evolved to a fully grammatical system before being overtaken by speech. But that's for chapter 6. First, we need to examine our ape ancestry for clues as to where language might have come from.

■ The grasping primate ■

Whatever Bishop Wilberforce thought of the matter, we are *primates*,[9] an order of mammals dating

[7] From The Devil's Dictionary (Bierce 1997). Ambrose Gwinnett Bierce's satirical definitions were originally compiled in The Cynic's Word Book, published in 1906. The reference given here is to the sixth edition.

[8] Márquez 1971, 11.

[9] It may be that the good bishop aspired to be a Primate, but he was evidently aghast at the thought that he was descended from primates. In the famous debate with Thomas Henry Huxley at the Oxford University Museum of Natural History in 1860, Samuel Wilberforce is said to have asked Huxley whether it was through his grandfather or his grandfather that he claimed descent from a monkey. Huxley is said to have replied along the following lines: "If then the question is put to me whether I would rather have a miserable ape for a grandfather or a man highly endowed by nature and possessed of great means of influence and yet employs these faculties and that influence for the mere purpose of introducing ridicule into a grave scientific discussion, I unhesitatingly affirm my preference for the ape." But we shall never know exactly what was said, since there is no verbatim account (Thomson 2000).

THE PRIMATE ORDER

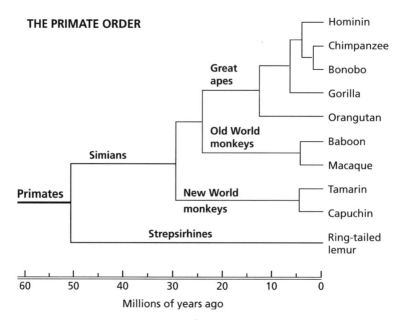

■ **Figure 3.1.** ■

A selection of living primates, showing the major transitions in their evolutionary history. The earliest known primate, *Purgatorius*, lived in eastern Montana some 65 million years ago. The Old World monkeys of Asia and Africa were separated from the New World monkeys of the Americas by continental drift and evolved separately from at least 30 million years ago, until humans bridged the gap. The great apes, which include the African apes (humans, chimpanzees, bonobos, and gorillas) and the Asian orangutan, date from about 15 million years ago. Humans are most closely related to chimpanzees and bonobos. Most primates are extinct; those shown are only a small sample of extant primates, selected because quite a lot is known of their behavior. The dates shown are approximate and subject to debate.

from some 60 million years ago (figure 3.1). Nearly all of the some 230 species of living primates spend most of their time in trees, although we humans are a conspicuous exception. Over the past 30 million years there has been a worldwide shrinkage of forests, which has meant that the primates have not thrived as they might otherwise have. Ironically, we humans have contributed to this, and the recent destruction of rain forests may well see the extermination of most of our primate cousins over the next century. Be that as it may, many of the characteristics of the primate order are adaptations to living in trees, and persist in modern humans.

The most distinctive bodily characteristic that primates share is the *hand*, which was adapted for grasping, with curling fingers and a thumb offset from the fingers to enable a closed grip, in which the tip of the thumb can touch the forefinger. The shoulders were also adapted to allow the arms to swing directly above the head, presumably to enable swinging from branches—an adaptation still exploited by trapeze artists and basketball players. In many primates, including our closest relatives, the chimpanzee and bonobo, the foot is also adapted to grasping, but this adaptation was lost to us after the emergence of upright walking in our forebears some 5 or 6 million years ago.

The arms and hands of primates are also well adapted to reaching and grasping objects of all sizes, at every position within an arm's length of the body and inside the field of view. Our primate heritage also allows us to reach behind the body, about as far with the outstretched arm as the eye can see when the head is turned. This kind of flexibility may have been important in climbing trees and swinging through the branches but probably owes something as well to food gathering, whether by plucking fruit or by catching insects. The hands and fingers are also specialized for the fine manipulation of objects.

Primates are also blessed with highly developed visual systems. The eyes are placed frontally, allowing for stereoscopic vision, and unlike other mammals primates see the world in color. It is reckoned that about half of the primate brain is involved with vision in one way or another,[10] and that the visual system is about as developed in the monkey as it is in humans. Indeed, much of what we know about *human* vision has come from tracking the brain circuits involved in vision in the monkey, typically by recording from single or multiple neurons in various parts of the brain while visual signals are displayed in front of the animal's eyes. Different areas of the monkey brain have been shown to be involved in different aspects of vision, such as the perception of color, movement, or even specific patterns, such as faces. Much of this work has been corroborated by modern imaging studies of the human brain; that is, the brain areas involved in human vision largely parallel those involved in monkey vision. We have gained in lots of ways over our primate forebears, notably in language and perhaps in an expanded awareness of time, and our brains have grown correspondingly bigger, but visually we are still creatures of the primeval forests.

[10] *Blakemore 1991.*

With highly developed control of the arms and hands and accurate three-dimensional vision, primates possess a natural basis for communicating about the world. Movements of the hands and arms can be controlled by the higher centers of the cerebral cortex, while vocalizations are largely (if not completely) controlled by more primitive subcortical areas. This means that hand movements can be *intentional*, flexibly programmed on-line, as it were, to respond to novel situations, whereas vocalizations are largely tied to fixed situations. We saw in the previous chapter how even the chimpanzee may be unable to produce sounds in the absence of the appropriate emotional state, or even to suppress sounds that are emotionally induced, just as humans are often unable to suppress laughter or crying.

Communication also requires the mapping of one's own bodily actions onto those perceived in others; in speech, for example, we need to understand that the words we utter are the same as the words uttered by others. Giacomo Rizzolatti, a neuroscientist working in Parma, Italy, has discovered that a mechanism for such a mapping appears to exist in the brains of macaque monkeys. He has recorded from single neurons (nerve cells) in the frontal lobe of the monkey brain that respond to particular reaching and grasping movements that the animal makes. These neurons are pretty fussy about what they respond to, so that individual neurons are specialized for particular movements. What is more remarkable, though, is that some of these neurons also respond when the monkey observes a human making the same movement that elicited the response when it was made by the monkey![11] Rizzolatti has called these neurons *mirror neurons*, because they seem to provide a mirror between action and perception. What you see is what you *do*.

These neurons were recorded in an area of the frontal cortex that appears to correspond to an area of the human brain involved in the production of speech—an area known as Broca's area after the nineteenth-century French physician Paul Broca, who discovered its role. This reinforces the speculation that mirror neurons constitute a precursor to language, which also requires a mapping between the production and the perception of complex actions. It will not have escaped the reader's attention, just as it did not escape Rizzolatti's, that the actions are manual, not vocal, strongly suggesting a gestural origin for language.[12] At some point, perhaps late in human evolution,

[11] *Rizzolatti, Fadiga, Gallese, and Fogassi 1996.*
[12] *Arbib and Rizzolatti 1997 elaborates his point.*

gesture yielded to vocalization, although Broca's area seems to play much the same role in signed language in the deaf as it does in spoken language in those who speak.[13] Evolution wrought another change: Broca's area is located in the left side of the brain in most people, whereas mirror neurons have been recorded on both sides of the macaque brain. As the programming became more complex, perhaps to incorporate syntax, it became lateralized in the brain. It is damage in the vicinity of Broca's area in humans that sometimes results in agrammatism—the condition in which speech is reduced to something like protolanguage, as we saw in chapter 2.

It is now clear that there is a mirror-neuron system in humans as well, and not only in those fluent in signed language. Rizzolatti and his colleagues measured brain activity with a technique called positron emission tomography (PET) and found that a number of areas, including Broca's area, were activated when people watched grasping movements made by others.[14] A more recent experiment using a technique called magnetoencephalography (MEG) recorded activation while people made a movement that involved reaching out and pinching the top of an object with the thumb and forefinger. There was activation in Broca's area on the left and in the so-called motor cortex on both sides. The same areas were activated by simply watching another person make the movement as by actually imitating the movement. MEG allows precise recording of *when* these areas were activated, and Broca's area was always activated first, followed by the motor cortex on the left, and then the motor cortex on the right. That is, Broca's area seemed to be in the driver's seat, organizing the actions, and also the *perceptions* of actions, that depend for their actual execution on the motor cortex.[15]

In addition, some evidence has recently emerged that Broca's area plays a critical role in the way humans organize remapping between movements of the hands and perceptions of those movements. A group of Japanese researchers have studied how people adapt to wearing prisms that reverse the way they see their hands, so that the left hand is seen as the right hand, and vice versa. This of course creates a discrepancy between vision and touch; the person may feel something with her left hand, but her eyes tell her that the right hand is touching the felt object. After a month of adaptation, people could even accomplish complex acts like riding a bicycle while wearing the

---

[13] *This is discussed more fully in chapter 6.*
[14] *Rizzolatti et al. 1996.*
[15] *Nishitani and Hari 2000.*

prisms.[16] Adaptation was also remarkably flexible in that, once adapted, the people could use either mapping at will. Using yet another technique known as functional magnetic resonance imaging (fMRI), the researchers measured the brain activity of these people when they wore the prisms. They found that Broca's area, on the left side of the brain, was activated regardless of which hand, left or right, was involved.[17]

These remarkable brain-imaging studies strongly suggest that Broca's area still plays a role in integrating hand movements with vision, a role that in humans has become confined to the left side of the brain but that has nothing to do with vocal language. One begins to wonder whether language is quite as unique as Chomsky and others have maintained. And, of course, the gestures made by Rizzolatti's monkeys also seem to have little relation to language, although they might well constitute a primitive kind of communication, in which the specific actions of one animal are matched to the same actions in another. Some evidence, in fact, indicates that the perception of speech works in roughly the same way; that is, people recognize speech sounds not so much by their acoustic properties as by how they are produced. This is known as the motor theory of speech perception.[18] But more significantly, mirror neurons suggest that the origins of expressive language may go back tens of millions of years, to some common primate ancestor, and may lie in visuo-manual adaptations rather than auditory-vocal ones. Language has been bequeathed to us, not by word of mouth, but as a hand-me-down.

■ The manipulative ape ■

Some 30 million years ago, a suborder of primates known as the apes became differentiated from monkeys.[19] Modern apes include gibbons, sometimes known as the lesser apes, and the so-called great apes, which include the orangutan, gorilla, chimpanzee, bonobo, and modern humans. Of course apes still share many features with the other primates, including some of the adaptations to life in the trees. But apes were (and are) larger

[16] I can remember as a child trying to ride a bike holding the handlebars with my arms crossed, with disastrous results. This is a bit what it is like when one wears reversing prisms.

[17] Sekiyama, Miyauchi, Imaruoka, Egusa, and Tashiro 2000.

[18] The motor theory has been modified and refined several times since the classic article by Liberman, Cooper, Shankweiler, and Studdert-Kennedy (1967). For a more recent and sophisticated statement, see Liberman and Whalen 2000.

[19] Or perhaps earlier. Two fossils discovered in Northern Egypt have been tentatively identified as apes, mainly on the basis of their dental features; these are Propliopithecus, dating from 33 to 35 million years ago, and Aegyptopithecus, dating from about 33 million years ago (Kay, Fleagle, and Symonds 1981).

than monkeys and so spent less time in the trees, gradually adapting to a more terrestrial existence—and thereby no longer hangs a tail. The gorilla is by far the largest of the modern great apes, large enough to be safe from most predators, and lives on a simple vegetarian diet.

One of the distinguishing characteristics of the apes is the flexible shoulder blade, allowing our fellow apes to swing from branches with the arm pointing straight upward—and allowing us humans not only to do that but also to indulge in such activities as serving in tennis, bowling in cricket, and swinging on the flying trapeze. Monkeys also swing nimbly from branches, but the arm is at an angle to the body instead of being aligned with it. But monkeys also walk easily on all fours, whereas the altered shoulder made this difficult in the great apes.[20] Chimpanzees and bonobos are sometimes able to walk upright, albeit in a very limited way, but for the most part they get around on all fours, as gorillas also do, supporting the upper part of their bodies on their knuckles, a mode of locomotion known as knuckle walking.

In apes, as in monkeys, the thumb can touch the forefinger, so that small objects can be held between these digits without touching the palm of the hand. Another development in the apes was a freeing of the forearm bone at the elbow, allowing the hand and lower arm to twist away from the thumb. If you place your hand with the palm facing up, you can then turn it right around by lifting the thumb and swinging it over, so that the palm is now downward. This is done entirely with a twist of the forearm at the elbow. You can then turn the hand a further 90 degrees, so that the thumb now points downward, but this requires a twist of the whole arm from the shoulder. In apes, therefore, the full available twist is 270 degrees, which greatly increases the range of grasping options.

The hands of the great apes are all rather alike, with thumbs well separated from the four fingers. As figure 3.2 shows, it is the gorilla hand that most closely resembles the human hand, despite the fact that the chimpanzee (and bonobo) are more closely related to humans. Yet the gorilla does not seem to spontaneously use tools, whereas the chimpanzee and bonobo have developed quite elaborate tool cultures. For example, different communities of chimpanzees have independently developed such varied tool cultures as fishing for

[20] *The nature and effects of this adaptation in apes are further discussed in Byrne 1995.*

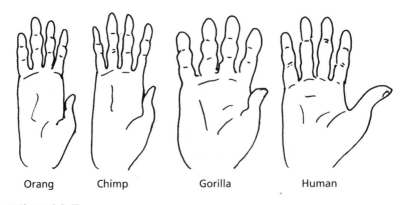

Orang   Chimp   Gorilla   Human

■ **Figure 3.2.** ■

The rise of the thumb: hands of the apes, scaled to the same length.

termites, ant dipping, nut cracking, and leaf sponging.[21] But we should not neglect the cognitive skill of the gorilla, who is in danger of being stereotyped, according to Richard Byrne, as "nice, but dull." Mountain gorillas, in particular, are highly skilled at preparing inhospitable leaves and other vegetation for their consumption. For example, when eating the nettle *Laportea alapites*, gorillas are highly dexterous at avoiding stings to the palms, fingers, and especially the lips. The bedstraw *Galium ruwenzoriense* is covered with tiny hooks, and the gorilla eats this with shearing bites rather than taking it through the lips, thereby avoiding the hooks becoming attached to the mouth. Byrne has constructed flow diagrams showing how gorillas gather and prepare different kinds of plants for eating. These include recursive components that somewhat resemble the recursive structure of language,[22] although they surely lack the flexibility and generativity of true language.

Other aspects of these skills are languagelike as well. They are fully established by the time the young gorilla is weaned, at three years old, just as the essence of human language is laid down by the time the child is about four. This suggests that the skills are learned either from the mother or from the male silverback leader, the only individuals who allow infants near them. Byrne argues, though, that the skills are not learned by strict imitation, since the way in which the infants actually accomplish the acts doesn't resemble the way their mentors do them any more than they resemble the way other adults

[21] *Reviewed in Tomasello and Call 1997.*
[22] *Byrne 1996.*

do them. Learning is more a matter of trial and error—reinforced by the mentor, perhaps—than of actual imitation. The two hands typically have complementary roles, and in about two-thirds of the animals observed, it was the right hand that assumed the role requiring the more precise manipulations. This suggests that the left side of the brain was dominant for fine movements, just as it is in the human brain, although the *proportion* of right-handers is considerably higher in the human population. It is also the left side of the human brain that is dominant for speech, but more of this in chapter 7.

It's not only the hands that are used in manipulation (despite the etymology of the term). Apes also have good voluntary control over the mouth and jaw and often use them for manipulation—usually, but not always, in the context of eating. This is one prerequisite for the production of articulate speech, but it is not the only one. As we shall see in chapter 7, breath control and the shape of the tongue and vocal tract had to change considerably before speech became possible in the evolution of our species.

■ The gesturing ape ■

In the previous chapter we saw that great apes have been taught quite successfully to communicate using manual gestures. They have also been observed to use gestures spontaneously, both in captivity and in the wild. Gestures typically occur in contexts with a clear social component, such as play, aggression, appeasement, eating, sex, and grooming.[23] Some gestures seem clearly to be intentional, since there is a flexible relation between the signal and its apparent purpose; unlike vocal calls, these signals are not fixed responses to fixed situations. For example, chimpanzees in the wild have been observed to use a single gesture for different purposes, such as an "arm-high" gesture for appeasement or to invite grooming, or different gestures for the same purpose, as when an infant chimpanzee either puts its arms around its mother's head or grabs her hand to induce her to continue playing. While gesturing in this fashion, chimps have also been observed to alternate gaze between the chimp they are signaling to and the goal. For example, a chimpanzee gesturing for food it cannot reach may look alternately at the food and at its mother.[24] After gesturing, chimps typically wait expectantly for a response, which is

[23] *Goodall 1986.*
[24] *See Plooij 1978 for this and other examples.*

further evidence that the gesture is intentional, designed to solicit a reaction.

The spontaneous gestures usually refer to actions rather than objects, and they are iconic rather than symbolic, which is to say that they mimic the actions the animals wish to communicate about. Nevertheless they are not completely iconic but may develop an increasingly abstract aspect with experience and social contact. This transformation from iconic to abstract may be termed *conventionalization*. Frans de Waal suggests that communicative gestures emerge from actions on the physical world and are then adapted and conventionalized.[25] Young chimpanzees quickly learn how to grab an object by the hand. Later, they may extend a grasping hand toward an object that is out of reach, indicating that they want it. Later still, they may extend an arm with the hand drooping downward toward a person, indicating that they want sympathy.

The progression from direct action to conventionalized gesture is actually a fairly general property of animal communication. Tom Givón gives the example of horses. Horses attack by turning their backs on their victims and kicking with their hind legs, lowering their heads and flattening their ears. In establishing a social hierarchy, this is reduced to display rather than actual attack, and the display is progressively reduced, so that once the hierarchy is established the flattening of the ears is sufficient to signal aggression.[26] The signals, therefore, become increasingly simplified and abstract, so that one really has to "know the code" in order to interpret them and take appropriate action. As we shall see in chapter 6, the process of conventionalization also occurs in the signed languages of the deaf.

Nevertheless, the conventionalization of communicative signals is undoubtedly much more advanced in humans than in chimpanzees. Terrence Deacon has dubbed us "the symbolic species," the only one with an intricate system of abstract symbols that can be manipulated independently of their referents.[27] Even the gestures of signed language have a strong symbolic element, which means again that one has to know the code in order to understand what they mean—although signed language is in general more iconic than spoken language, as we shall see in chapter 6. In contrast, Wolfgang Köhler, in his studies of chimpanzees in the Canary Islands, observed that most of the gestures the animals made were imitations of desired

---

[25] *de Waal 1982.*
[26] *Givón 1995.*
[27] *Deacon 1997.*

actions.[28] For example, if one chimpanzee wanted to be accompanied by another, he gave her a nudge, or pulled her hand, and mimed walking. One chimpanzee who wanted to be petted stretched her arm out to the people present and awkwardly stroked and patted herself, at the same time gazing pleadingly.

Joanne Tanner and Richard Byrne have also counted some thirty different gestures made by lowland gorillas in the San Francisco Zoo, where the animals are enclosed in a large, naturalistic area. Detailed analysis of a selection of these gestures revealed them to be largely iconic, and easily understood by both human and gorilla observers.[29] Kanzi, the bonobo whose linguistic skills were described in the previous chapter, has also developed many iconic gestures. He uses hitting motions to indicate that he wants nuts cracked, twisting motions with his hands to indicate that he wants a jar opened. Bonobos are also reported to use hand and arm gestures to other bonobos to show the positions they would like them to adopt for copulation (bonobos are famously sexy).[30] It may well be that the ability of chimpanzees and gorillas to both produce and interpret gestures depends on a mirror-neuron system similar to that in monkeys.

To give you some impression of the variety of chimpanzee gestures, table 3.1 lists thirty gestures that were observed in a study by Michael Tomasello and his colleagues on free-ranging chimpanzees at the Yerkes Regional Primate Center Field Station in Atlanta, Georgia.[31] They by no means exhaust the full repertoire but were chosen because they can be readily observed and tabulated by mere humans. Notice that nearly all of them include reference to "the other." That is, they are *dyadic*, involving an interaction with one other individual, usually in a way that invites reciprocation. Gestures are produced more often when the recipient is looking, indicating that the gesturer is sensitive to when others are watching them. The dyadic nature of these gestures distinguishes them from chimpanzee *vocalizations*, which are not directed to specific others. In this sense, at least, chimpanzee gestures are more like language than are their vocalizations.

Unlike the gestures of human children, though, chimpanzee gestures are seldom *triadic*. A triadic gesture is one that involves a third object in addition to the gesturer and the intended recipient. From an early age human children will gesture to indicate objects that are at a distance from themselves.

[28] *Köhler 1925.*
[29] *Tanner and Byrne 1996.*
[30] *These examples are from Savage-Rumbaugh and Lewin 1994.*
[31] *Tomasello, Call, Warren, Frost, Carpenter, and Nagell 1997.*

■ **Table 3.1** ■
A sampling of chimpanzee gestures

Arm-on	Subject approaches the other with its arm extended and places its arm on the other's back
Arm-raise	Subject raises its arm (as if to hit), often before charging
Back-offer	Subject insistently puts its own back in the face of the other
Ball-offer	Subject presents ball to the other, taking it back to invite wrestling
Belly-offer	Subject presents its belly to the other
Direct-hand	Subject puts the other's hand in its own arms
Foot-stomp	Subject puffs out its chest and approaches the other upright, stomping
Genital offer	Subject leans back and frontally presents its genitals to the other
Ground slap	Subject slaps the ground (or an object) and looks to the other
Hand-beg	Subject places its hand under the other's mouth and looks in the face of the other
Hand-clap	Subject slaps its own wrist or hand and approaches the other
Head-bob	Subject "bobs and weaves" in bowing position at the other
Head-shake	Subject rapidly shakes head horizontally at the other
Lead	Subject pulls the other along by the scruff of the other's neck
Leg-offer	Subject puts its leg in front of the face of the other and "tries" to run away
Lip-lock	Subject sucks the other's lower lip and then backs away
Look-back	Subject runs away, looking over its shoulder at the other
Point	Subject points to its own side and looks to the other's face
Poke-at	Subject pokes a body part of the other
Push-object	Subject pushes an object in the other's direction
Raise-object	Subject holds an object above its head
Reach	Subject extends an arm to the other
Rub-chin	Subject strokes chin of adult and looks to its face
Shake-object	Subject holds an object and pushes it back and forth
Spit-at	Subject spits water toward the other
Swagger	Subject stands, usually on two legs, and rocks from side to side
Throw-stuff	Subject throws some loose material at the other
Touch-side	Subject touches the side of the other
Wave-object	Subject swings an object horizontally either in front of it or over its head
Wrist-offer	Subject cautiously extends the back of its flexed wrist to the other

Moreover, the individuals they gesture to are also often at a distance, whereas chimps typically gesture through direct physical contact.[32] An exception, though, is pointing. As we shall see below, chimpanzees can be taught to point in triadic fashion to objects they cannot reach.

Tomasello's work shows that chimpanzees gesture in markedly individual ways, suggesting that at least some of the gestures are learned. Chimpanzees within the same group or generation tended to use more gestures in common than those in different groups or generations. This is also true of human language. Isolated groups develop dialects that eventually evolve into different languages, and each generation seems to invent new words that their elders do not understand or refuse to acknowledge. We don't understand our kids.

Tomasello argues, though, that chimpanzee gestures are not acquired through imitation. When the human experimenters taught selected chimpanzees new gestures outside their groups and then reintroduced them to the groups, the others showed no tendency to copy. Tomasello suggests that gestures are acquired through emulation, not imitation.[33] That is, the gestures are reinforced by others, and this induces some uniformity, but in no case does actual copying seem to occur. Human children, by contrast, readily imitate the gestures of others, just as they readily pick up new terms from television or pop stars. Remember, too, from the previous chapter that chimpanzees do not seem to imitate when solving mechanical problems. Parrots may parrot, but apes don't ape.[34]

Gestures often involve movements of the mouth and face, including the use of expressions that resemble human expressions. Sometimes, though, the resemblance can be misleading. Humans often smile with the teeth showing, but when chimpanzees bare their teeth, it is advisable to keep out of the way. Of course, facial expressions are often emotional rather than intentional, and it is sometimes hard to tell the difference. Nevertheless there can be little doubt that the face is part of the gestural system. As we shall see in chapter 6, facial expression is also critically involved in the signed languages of the deaf.

[32] Tomasello and Camaioni 1997. These authors also observed that the autistic children's gestures were more like those of chimpanzees than those of normal children.

[33] This may seem curious, given the existence of mirror neurons, which we must presume to be present in the chimpanzee as in the macaque brain. Mirror neurons would seem to provide a natural mechanism for imitation.

[34] This has become something of a controversial issue. Some authors have argued that apes can emulate but not imitate. Emulation is defined as reproducing the end result of an action, without copying the exact movements required to get there. It has been claimed that chimpanzees can imitate when learning to use tools from their fellow chimps (Greenfield, Maynard, Boehm, and Schmidtling 2000). But the distinction between emulation and imitation is often a fine one, and one could make the case that emulation requires more sophisticated cognitive processing than imitation. For further discussion see Byrne and Russon 1998 and ensuing commentaries.

■ What's the point? ■

A gesture of special interest is *pointing*, which offers one solution to the problem of reference. That is, we can indicate what a word means by pointing to its referent. In fact young children point before they speak[35] and continue to refer to things by pointing at them until they learn the appropriate words.[36] Pointing may therefore be critical to learning words and sometimes continues until discouraged by parents, many of whom consider it rude.[37] "Don't point, dear," my mother used to say, "it's not nice." We might sympathize with Shakespeare's Othello, wrongly believing himself humiliated by the faithlessness of his wife, Desdemona:

> but, alas, to make me
> A fixed figure for the time of scorn
> To point his slow unmoving finger at . . .

Old World monkeys, more polite than we are, have never been observed to point in the wild, but it appears that they are readily corrupted. Experiments have shown that monkeys can be taught to move an arm or the fingers into alignment with a visually perceived target, and that this can be accomplished without either seeing the pointing arm or receiving kinesthetic signals from it.[38] In this respect, pointing is like speech, which also occurs without feedback. No one has seen the great apes pointing in the wild, either, although Wolfgang Köhler, who studied chimpanzees in the Canary Islands while interned during World War I, observed that many of their gestures were transitional between grasping and pointing. The great Russian psychologist Lev Vygotsky royally remarked, "We consider this transitional gesture a most important step from unadulterated affective expression toward objective language."[39]

It appears to be a step easily taken. All four species of great ape—orangutans, gorillas, chimpanzees, and bonobos—have been taught to point by humans, typically as part of training to communicate with a form of signed language.[40] At first, pointing is taught as a means of indicating which individual is referred to, or of indicating particular keys on a keyboard containing visual symbols, but

[35] Iverson, Capirci, and Caselli 1994.
[36] Acredolo and Goodwyn 1988.
[37] In some groups of indigenous people in western North America, according to Hewes 1981, it is forbidden to point at rainbows.
[38] Taub, Goldberg, and Taub 1975.
[39] Vygotsky 1962, 35.
[40] For chimpanzees, see Savage-Rumbaugh 1986; for bonobos, see Savage-Rumbaugh, McDonald, Sevcik, Hopkins, and Rubert 1986; for gorillas, see Patterson 1978; and for orangutans, see Miles 1990.

in all cases the apes have spontaneously begun to point to other objects they want to have or to places they want to visit. In these cases, pointing is triadic, since it involves a third object or location besides the gesturer or the recipient. That is, the ape is now communicating about something *else*, which is an important advance toward language.

Although chimpanzees in the wild seem to live pointless lives, captive chimpanzees are quick to get the point, even without being explicitly taught by humans. In a study of 115 chimpanzees housed at the Yerkes Center, half a banana was placed outside their cages just out of reach.[41] None of these chimpanzees had been given language training, and only three of them had been previously taught to point. Yet fifty-three of them spontaneously pointed at the banana, and all of them alternated their gaze between the banana and the human experimenter. It is unlikely again that they learned to point by imitating humans, since only six of them pointed with the index finger, as humans do. The rest pointed with the whole arm and hand. It is also of interest that of those chimpanzees using a one-handed gesture, two-thirds used the right hand.[42]

One form of pointing that does seem to come naturally to the chimpanzee is eye gaze. Simply looking at something can cause other people to look at it too, as you can easily discover by looking upward, even at nothing in particular: other people in the vicinity will soon be following your gaze. Chimpanzees also naturally and easily follow the gaze of others, and human children show this ability as early as the second year of life.[43] But there is evidence that chimpanzees do not interpret or understand pointing or eye gaze in quite the same way that humans, even three-year-old humans.

This was illustrated by experiments carried out by Daniel Povinelli and his colleagues.[44] Chimpanzees can easily be taught to approach people they know and beg for food. If a person sits in front of a chimpanzee and points to one of two boxes, on the left or on the right, the chimpanzee understands readily enough that if it wants food, it should go to the box that the person is pointing to. But the choice breaks down if the person points from some distance away, and is systematically reversed if the person sits closer to the box that does not contain the food and points to the other one. It

[41] *Leavens and Hopkins 1998.*
[42] *Hopkins and Leavens 1998.*
[43] *Moore and Corkum 1998; Tomasello, Hare, and Agnetta 1999.*
[44] *These clever experiments are summarized and discussed in Povinelli, Bering, and Giambrone 2000.*

seems that chimps respond on the basis of how close the pointing hand is to the box containing the food, and not on the basis of where the hand is actually pointing. Again, young children have little difficulty interpreting a pointing gesture.

Similarly, when confronted with a choice between two people, one with and the other without a blindfold over her eyes, chimpanzees do not seem to appreciate that they should beg from the one who can see. (Perhaps it's just that beggars can't be choosers.) The same is true when one of the people has a bucket over her head or is covering her eyes with her hands. Only when one of the people is actually facing away from the animal does the chimpanzee easily choose the one facing toward itself. Young children, on the other hand, quickly recognize that they should approach the person who can see them, and that this depends on the eyes! The failure of the chimpanzee to appreciate this does not arise from failure to observe the eyes, since they readily follow the gaze of a person confronting them. Chimpanzees may eventually choose the person who can see them, but this is probably based simply on painstaking associative learning, and not on the understanding that eyes are for seeing.

One may be tempted to conclude from this that chimpanzees are simply rather stupid and have to rely on dogged learning rather than on "theory of mind," the understanding that others have mental states like one's own. Povinelli argues, though, that many behaviors, like following eye gaze, have the same basis in humans as in other primates, but that we "reinterpret" these behaviors as being more sophisticated than they really are. For example, people may spontaneously follow the gaze of someone who seems to be looking at something in the sky without reasoning along the lines of "That fellow must see something interesting up there." Gaze following may simply be an adaptive response that alerts other animals to danger or reward, but we humans have intellectualized it, often after the fact. Indeed, much of what we call introspection, or the sense that we have consciously "made up our minds" about something, may actually be rationalizations for behavior that was induced by automatic processes we are unaware of.

Nevertheless, intellectualization no doubt has benefits, or it would not have evolved. It may sharpen our communicative skills and bypass the need for associative learning. Once we have discovered the principle of how something works, there is no need for further learning; if children know that without eyes there is no seeing, they can use this knowledge in a wide variety of situations, including sneaking a cookie when no one is looking. If Povinelli is correct, chimpanzees do not seem to advance beyond simple associative learning. In the previous chapter, I recounted how he also taught chimpanzees to use hooked tools to haul in a wooden platform with a banana on it, but they could not adapt this solution when the task was changed slightly. That is, they apparently did not learn a *principle*. The human ability to project into the minds of others may be another example of a principle that chimpanzees do not grasp, a principle that enhances the ability to interpret other people's actions and modify one's own behavior accordingly. But perhaps this ability was also crucial for that other capacity that seems to be uniquely human—language!

We should perhaps not judge the chimpanzee too harshly when it comes to communicating with the eyes, since we humans are especially good at it. This is evident in the actual structure of the eye. We are exceptional among primates in having eyes in which the sclera, the coat of the eyeball, is white rather than pigmented, and much more of it is visible in humans than in other primates. The human eye is also exceptionally elongated horizontally.[45] The dark color of the exposed sclera in nonhuman primates may be an adaptation to conceal the direction of eye gaze from other primates or predators, whereas the human eye seems to have evolved to enhance communication rather than to conceal it—further evidence, perhaps, of the role of gesture in the evolution of language even in our hominin ancestors. Women, perhaps, understand the power of the eye better than men do, for as Biron observed in *Love's Labour's Lost*:

> From women's eyes this doctrine I derive:
> They sparkle still the right Promethean fire;
> They are the books, the arts, the academes,
> That show, contain, and nourish all the world;
> Else none at all in aught proves excellent.[46]

But whether or not their eyes have it, the chimpanzee may have been somewhat slighted by Povinelli's work. By nature, chimpanzees are competitive creatures, and one may wonder why they

[45] Kobayashi and Kohshima 2001.

[46] Act 4, scene 3. This is not to say that men are incapable of using their eyes to convey messages and influence the world. In P. G. Wodehouse's The Code of the Woosters, one Roderick Spode is described as a "Big chap with a small moustache and the sort of eye that can open an oyster at sixty paces."

should believe what humans are trying to tell them. Conversely, we humans are pretty competitive too and have plenty of incentive to show that apes are dumber than we are. Dogs, in contrast, have been bred to cooperate with humans, and Brian Hare has shown that dogs do seem to be able to locate a food source by observing where a person or another dog is looking or pointing.[47] He has also shown that chimpanzees are aware of what other *chimpanzees* can see, and that they modify their behavior accordingly. For example, a chimpanzee will approach food when a more dominant chimpanzee cannot see the food, but will be reluctant to do so when they can see that the dominant chimpanzee is watching![48]

Further, it is remarkable how well great apes like Kanzi, the bonobo whose linguistic skills were described in the previous chapter, can pick up communicative skills that they apparently do not use in the wild. Kanzi and other language-trained apes have lived and worked closely with their human mentors and are perhaps unusually willing to cooperate with them. Whether in problem solving or in communication, chimpanzees and bonobos probably do have the ability to make representations, to use symbols to stand for those representations, and to discover appropriate combinations through trial and error. But we have little evidence that they can go beyond associative thinking to derive *rules* of the sort needed to build grammar. As we saw in the previous chapter, many psychologists, such as Steven Pinker, have adopted the Chomskyan view that language is a highly specialized biological adaptation, with properties that are not dependent on general intelligence. But the crucial adaptation may not have been language itself, but the ability to think recursively, and so project into the minds of others. As I noted in chapter 1, the structure of the understanding that "I know that he can see me" has the same recursion as the sentence that expresses it.

Chimpanzees may be capable of at least limited recursion. Patrizia Potì, an Italian psychologist, has studied the way chimpanzees spontaneously organize objects, such as cylinders, square rings, crosses, and sticks, into groups. As young children do, they spontaneously group the objects around them and then perform operations on the groups. For example, they might take a pair of rings and rearrange them by placing one on top of the other; or they might make a set consisting of three objects, then take one object away and re-

---

[47] *Hare and Tomasello 1999.*
[48] *Hare, Call, Agnetta, and Tomasello 2000.*

place it with another. She calls these operations on a single group *first-order operations.* She also identifies *second-order operations,* which are effectively operations on operations. A simple example is the combining of two sets to form one bigger set, which may then be divided again into two smaller sets, and possibly rearranged in the process. According to Poti, macaques and capuchins do not indulge in second-order operations, and therefore do not demonstrate recursion. Chimpanzees do, but lag well behind human children in their capacity for it and, as far as we know, never progress beyond second-order operations.[49]

Perhaps it is in the extended use of recursion that humans truly excel. By the age of four or so, human children appear able to grasp the concept of recursion well enough to apply it repeatedly. We adults can understand sentences like *I suspect that she knows that I'm watching her talking to him,* which means that we can not only create and parse sentences like this but also appreciate the situations to which they refer. As I noted in chapter 1, recursion may be limited in practice by constraints on short-term memory, but the *concept* is an open one, and we can in principle apply it as often as we want. If not present in the chimpanzee or bonobo, this capacity was probably not present in the common ancestor we share with these animals and must therefore have emerged at some later point in the evolution of our species.

■ The cultured ape ■

Another characteristic that we humans like to claim as our own is what we are pleased to call *culture,* although not all are agreed that culture is a good thing. The English architect and planner Lord Esher remarked in the House of Lords, "When politicians and civil servants hear the word 'culture' they feel for their blue pencils."[50] The term *culture* has a variety of meanings, but here we may take it to refer to differences in habits, beliefs, and practices between communities. Even language is partly cultural, since the actual languages we speak are acquired from our parents or the communities in which we live, although the capacity to acquire language is a biological endowment. Culture includes religion, the

[49] *Poti 1997.*
[50] *In case you weren't there, or were there and missed it, this remark appeared in a speech he made on 2 March 1960. The good lord was probably improvising from a remark attributed to Hermann Goering, often quoted as "Whenever I hear the word 'culture,' I reach for my pistol." But probably behind both versions lies a line from Hanns Johst's 1933 play* Schlageter: *"Whenever I hear the word culture . . . I release the safety-catch of my Browning!" (act I, scene 1).*

style of clothes we wear, the way we do our hair (if we still have any), and so on. But culture is not wholly restricted to our own species. Other primates and, of course, birds also show cultural variations. For example, Japanese macaques at Koshima are unique among that species in that they wash sweet potatoes before eating them.[51] I suppose one might argue that those who washed their potatoes had a better chance of survival, which led to selection of the potato-washing gene, but more likely potato washing was an activity discovered by an enterprising member of the group, and others slavishly followed suit. Such is the nature of fashion.

The cultural diversity among chimpanzees, however, is greater than that so far recorded in any other nonhuman species—further evidence, perhaps, that the chimp may not be quite the chump that some have supposed. Andrew Whiten and his colleagues, examining the evidence from six different chimpanzee communities, have identified thirty-nine different behavioral patterns where the differences between communities could not be attributed to differences in physical or geographic conditions.[52] These patterns include grooming, courtship, and the use of tools. For example, nut cracking is found only in the two western communities (Taï Forest and Bossou) and not in the four eastern ones (Budongo, Gombe, Kibale, and Mahale). Chimpanzees at several sites use sticks to dip for ants, but they use different methods at different sites. At Gombe they hold a long wand in one hand and wipe the ants off it with the other, while at Bossou and in the Taï Forest they use a shorter stick and transfer the ants directly to their mouths. (All of the behavior patterns catalogued by Whiten and his colleagues involve *actions*, not vocalizations. We saw in the previous chapter that there may also be cultural variations in the pant-hoot call of chimpanzees, but it seems clear that chimpanzees deal primarily in a world of seeing and doing, where actions speak louder than words.)

Chimpanzee parents also seem to help their children learn in much the same ways that humans do. In the Taï forest, when a mother goes to gather more nuts, she leaves a nut positioned in the hollow of a tree root, which serves as an anvil, and places a hammer stone on top of it, so that her infant can pound the nut in her absence. This kind of maternal help is known as *scaffolding*. In a more complex example, a mother was seen to watch while

---

[51] Imanishi 1957.
[52] Whiten, Goodall, McGrew, Nishida, Reynolds, Sugiyama, Tutin, Wrangham, and Boesch 1999.

the infant attempted the action and intervened to clean the anvil and reposition the nut before the infant tried again to strike it. Curiously enough, chimpanzees at the Gombe National Park in Tanzania do not use scaffolding when they transmit information to their infants on how to fish for termites. However, the young chimpanzees closely observe their elders and indeed appear to imitate the action rather closely—contrary to the claim that chimpanzees can't imitate. In short, chimpanzee tool technology is transmitted through what might be termed apprenticeship.[53]

Of course, human culture is much more varied, and the forms of apprenticeship much more elaborate, including that much-loved institution, the school. Even so, we can discern in chimpanzee society at least the precursors of our own cultural diversity. To some extent, the diversity of human cultures and human technological sophistication can be attributed to a ratchet effect. New technologies build on old ones, and the transmission of knowledge between generations allows what seems like a never-ending progression—to the point where it is difficult to cope with the changes that occur in a single lifetime. But biology must have played its part in our stunning capacity to create a transmitting culture, including the most efficient means of cultural transmission so far invented. And that, of course, is language.

■ The gestural theory of language evolution ■

In this chapter I have tried to lay the foundation for the view that human language evolved from gestures of the hands and face, rather than from primate vocalization. Primate calls are largely emotional, and tied to specific situations, such as danger, sexual engagement, or the discovery of food. It is not surprising, therefore, that we have had virtually no success in teaching even our closest relatives, the chimpanzees and bonobos, to talk. Primate gestures are an entirely different matter. Primates have evolved hands that are capable of a wide variety of actions, and their hands and arms are under precise cortical control. This is why we *have* been able to teach apes to communicate using gesture, at least to the level of protolanguage, as we saw in the previous chapter. And it is becoming clear that they use communicative gestures in the wild, often in one-on-one situations that somewhat resemble human conversational language. Primate calls, by con-

[53] *See Greenfield et al. 2000 for a useful review.*

trast, are typically directed to the community at large, and to no one in particular.

The idea that language may have evolved from gesture is not new. As I shall elaborate in chapter 7, the eighteenth-century philosopher Condillac was one of the first to enounce this idea, and my own account of how vocal language eventually came to dominate is scarcely any improvement on his. Similar themes were pursued in the nineteenth century. Even Charles Darwin made some concession to the role of gestures: "I cannot doubt that language owes its origins to the imitation and modification of various natural sounds, and man's own distinctive cries, *aided by signs and gestures*" (emphasis added).[54] Wilhelm Wundt, who founded the first laboratory of experimental psychology at Leipzig in 1879, referred to "the assumption, outspokenly held by many anthropologists, that gestural language is the original means of communication."[55] Wundt is often discredited for having tried to found experimental psychology on introspection, the subjective "looking inward" on the mind, whereas modern experimental psychology relies on objective data, what people actually do, so that mental processes are based on inference rather than on direct observation. Yet Wundt's analysis of gestural communication is in many ways remarkably modern, although he underestimated the linguistic sophistication of the signed languages invented by the deaf. According to Wundt, humans share a number of basic expressive gestures with other animals, but the "great step" that set humans apart was the ability to imitate "arbitrary activities."[56] In the hands of humans, as it were, gesturing acquired at least some of the properties of true language.

The British neurologist MacDonald Critchley was greatly interested in gesture. He lamented that the publication of his book *The Language of Gesture* coincided with the outbreak of World War II,[57] and was therefore largely ignored, so he wrote a second book, *Silent Language*, which was published in 1975. "Gesture," he wrote, "is full of eloquence to the sagacious and vigilant onlooker who, holding the key to its interpretation, knows how and what to observe."[58] Critchley speculated that gesture might be the precursor of speech but declared himself unable to accept that the earliest human language was purely gestural

[54] Darwin 1896, 87.
[55] Wundt 1973. This book is a translation of chapter 2 of the fourth edition (1921) of Wundt's Volkerpsychologie, first published in 1900.
[56] Wundt 1921, 128.
[57] It is not clear which he considered the more lamentable, the neglect of his book or the outbreak of war.
[58] Critchley 1975, 221.

and voiceless. But he then seemed to contradict himself, going on to argue that gesture predated speech as a mode of communication in human evolution.

The gestural theory was given comprehensive treatment by the anthropologist Gordon W. Hewes, notably in a 1973 article in the journal *Current Anthropology* in which he touched on most of the basic arguments covered in this book. Given the luxury of space, I have embellished these arguments, have added points arising from more recent research, and in chapter 9 have tried to deal with one question that Hewes had difficulty answering: Why was gesture eventually superseded by speech? But the credit for the gestural theory in its modern form really belongs largely to Hewes, and certainly not to me.

Another major influence on the field was the late William C. Stokoe, who died early in 2000, just as I began this book. Stokoe was largely responsible for reintroducing American Sign Language as the official language at Gallaudet University, in Washington, D.C., but he was also influential in persuading linguists that natural signed languages like ASL are true languages and not mere surrogates. The 1994 book *Gesture and the Nature of Language,* which Stokoe coauthored with the physical anthropologist David F. Armstrong and the linguist Sherman E. Wilcox, contributed greatly to the theory that language itself originated in manual gestures. This theme is further advanced, from the point of view of signed language, in David F. Armstrong's *Original Signs* (1999). I discuss the nature and relevance of signed languages in chapter 6.

4 ■ On Our Own Two Feet ■

The Sphinx was a monster of Greek myth.[1] She had a woman's head and breasts, the body and paws of a lion, wings, and a serpent's tail. She lived near Thebes and terrorized the inhabitants by setting them riddles and devouring those who could not solve them. According to the oracles, one of her riddles was so difficult that the Sphinx would kill herself if it were ever solved. Here is the famous "riddle of the Sphinx":

> What goes on four feet, on two feet, and three,
> But the more feet it goes on, the weaker it be?

It was eventually Oedipus who came up with the answer: a human being! In infancy, we crawl of all fours, while in old age we use a stick to help prop up our failing legs. Only in the prime of life do we walk erect, on two feet. The Sphinx duly killed herself, and the good people of Thebes were delivered from her tyranny. Oedipus, though, seems to have stayed around to cause trouble, at least in our fantasies.

Upright walking, or bipedalism, is the main feature that distinguished humans and our forebears from the other great apes (chimpanzees, bonobos, gorillas, and orangutans). If language was built on gesture rather than on vocal calls, then bipedalism could have been an important step, so to speak, toward language, since it would have

[1] This Sphinx is not to be confused with the other one, the huge statue near the Great Pyramids at El Giza in Egypt.

freed the hands and arms from involvement in locomotion and thus allowed gesture to develop more freely. More than that, while freeing the hands, it would have enslaved the feet! In other primates, including the chimpanzee, the feet are effectively a second pair of hands, capable of grasping and manipulating objects, whereas the role of human feet is effectively just to bear weight. To be sure, some people born without arms have managed to use their feet as substitute hands, even writing or drawing with them. But in most of us, the toes are more or less useless appendages, merely reminders of our arboreal past. What this means is that the area of the brain responsible for initiating movements of the feet and toes is much reduced relative to the corresponding area in the monkey or chimpanzee brain, allowing more neural space for control of the hands. Moreover, the areas in the so-called motor strip are dependent on experience, so that the more you use your hands and the less you wiggle your toes the more neural space is dedicated to the hands and the less to the feet.[2]

The rallying cry of the ascendant pigs in George Orwell's *Animal Farm* was "Two legs bad, four legs good!" but for our hominin ancestors it was the other way round. Let's therefore examine the evolution of bipedalism more closely and ask how and why it may have come about.

■ Splitting from the apes ■

About 6 million years ago, there existed a single species that was the common ancestor of ourselves, the modern chimpanzee, and the modern bonobo. This species split into two branches, with one branch later splitting into the chimpanzee and bonobo species and the other branch giving rise to a number of different species, among which we are pleased to number ourselves. These species, who comprise the family known as *hominins*, can actually be classified into seven different *genera*, each containing one or more *species*, all but one of which eventually became extinct.[3] Classification of these species is still somewhat in dispute, but one recent taxonomy, embellished with even more recent discoveries, is shown in table 4.1.[4] Were it not for the sur-

[2] Richards 1986.

[3] The terminology can be a little confusing. We belong to the primate order, the hominin family, the genus Homo, and the species Homo sapiens. Oh, and genera is the plural of genus.

[4] This is based on a recent reclassification by Wood and Collard 1999, who admit that this taxonomy may err on the side of being too "speciose." But since then, the situation has grown worse, and a little further on in this chapter you will meet two further species, only recently (and controversially) identified, called Kenyanthropus platyops and Orrorin tugensis.

■ Table 4.1 ■
Classification of the hominins

Genus	Species
Orrorin	*Orrorin tugensis* *
Ardipithecus	*Ardipithecus ramidus*
Australopithecus	*Australopithecus anamensis*
	Australopithecus afarensis
	Australopithecus bahrelghazali *
	Australopithecus garhi
Kenyanthropus	*Kenyanthropus platyops* *
Praeanthropus	*Praeanthropus africanus*
Paranthropus	*Paranthropus aethiopicus*
	Paranthropus boisei
	Paranthropus robustus
Homo	*Homo rudolfensis* **
	Homo habilis
	Homo ergaster
	Homo erectus
	Homo antecessor *
	Homo heidelbergensis
	Homo neanderthalensis
	Homo sapiens

*Added to the taxonomy presented in Wood and Collard 1999.
**Perhaps better classified as *Kenyanthropus rudolfensis*.

vival of the one remaining species, *Homo sapiens*, this book could not have been written.

As far as we know, all of the species identified as hominins have been bipedal; indeed, if they were not, we would probably not welcome them into the hominin family. Modern chimpanzees, bonobos, and gorillas can stand upright, and even walk bipedally in a limited fashion, but their main method of getting around on open land is a form of quadrupedal ("four-footed") locomotion known as knuckle walking, in which the knuckles of the hands make contact with the ground. Bipedality may therefore be considered a family characteristic, and perhaps the main defining one, of the *Hominini*, or hominins.

Very little was known about that 5 or 6 million years of transition from ape to human until 1924, when a young anatomist called Raymond Dart came into possession of a skull with both humanlike and apelike features that had been found in a cave near Taung, South Africa. He called it *Australopithecus africanus*. The term *australopithecus* has nothing to do with the fact that Dart was an Australian; it simply means "southern man." Dart published the news in the prestigious scientific journal *Nature* in 1925, hailing his find as the "missing link." He was at first ridiculed by the scientific establishment, but subsequent events proved him right.[5] His discovery opened the floodgates, and since then scores of other hominin fossils have been discovered in southern and eastern Africa. What used to be called the missing link has become a tangled web of perhaps twenty different species, all but one of which eventually became extinct. In terms of survival, the hominin episode in evolutionary history was not very successful, but we can console ourselves with the knowledge that the remaining hominin is ourselves.

The earliest creature tentatively identified as a hominin dates from around 6 million years ago. This is *Orrorin tugensis*, whose fossil remains were discovered in Tugen Hills in Kenya; he is the most recent claimant to the title of grandfather of us all.[6] *Orrorin* means "original man" in the local Tugen language. *Orrorin*'s claim to hominin status is controversial, in part because its age of 6 million years stretches the limits of molecular estimates as to when the common ancestor of ourselves and the chimpanzee lumbered around on African soil.[7] Nevertheless, the age of the fossil does not seem to be in doubt, and examination of the bones seems to show that *Orrorin* was bipedal when on the ground but also retained some adaptation to life in the trees.

The next oldest hominin so far discovered dates from some 4.4 million years ago and is less controversial.[8] It has been cautiously named *Ardipithecus ramidus*—*Ardipithecus* rather than *Australopithecus* because it is still not entirely clear whether it was a bipedal hominin or a quadrupedal ape, and *ramidus*, which is the Latin for "root," because it lies very close to the common ancestor of the hominins and chimpanzees, notwithstanding the claims made more recently for *Orrorin*.[9] A slightly later hominin, dating from about 4.2 million years ago and known as *Australopithecus anamensis*, was almost certainly bipedal.[10] A hominin previously known as *Australopithecus afarensis*, but now reclassified as *Praeanthropus africanus* (see

[5] It is actually a little misleading to call A. africanus a "missing link," since this implies it was neither ape nor human but transitional between the two. The australopithecines were great apes that had adapted to the conditions of southern and eastern Africa about as well as chimpanzees and gorillas have adapted to the forests of western Africa. The australopithecines lasted at least 3 million years before becoming extinct. We should be so lucky.

[6] Senut, Pickford, Gommery, Mein, Cheboi, and Coppens 2001. For a cautious appraisal of this claim, see Aiello and Collard 2001.

[7] The controversy is compounded by the fact that Martin Pickford, one of the authors of the article describing Orrorin, has been at odds with the renowned Leakey family who have long dominated paleontological research in Kenya, and in fact was at one point arrested for collecting without a permit (see Butler [2001] for more details). It is all a bit of an orror story.

[8] White, Suwa, and Asfaw 1994.

[9] White, Suwa, and Asfaw 1995.

[10] Leakey, Feibel, McDougall, and Walker 1995.

table 4.1), dates from some 3.2 million years ago. This species includes the famous Lucy from the Hadar region of East Africa and was also bipedal,[11] although recent evidence from the bones of the wrist in both *Praeanthropus africanus* and *Australopithecus anamensis* points to an earlier knuckle-walking phase.[12] This suggests that the common ancestor of ourselves, the chimpanzee and bonobo, and probably also the gorilla, was a knuckle-walking ape.

Very recently, a possible companion for Lucy has turned up, also dating from around 3 to 3.5 million years ago. This species, whose fossil remains were found near Lake Turkana in Kenya, has been named *Kenyanthropus platyops*.[13] It is distinguished from *Australopithecus afarensis* by a peculiarly flat face and small teeth, and in these respects is more like a modern human than the other early hominins, although its brain case is no larger than a chimpanzee's. So different is it from *afarensis* that its discoverers have suggested that it belongs to a separate genus, bringing to seven the number of different proposed genera of hominins, and to nineteen the number of proposed species, as shown in table 4.1. Some would consider this classification extravagant, and some further reclassification is likely as more specimens are discovered, and as present-day hominins look back on their extinct ancestors and sort through their bones.

These numerous species, with their complicated names, are bound to cause some confusion, although I suppose it is as well to get to know the family. Figure 4.1 shows approximately when the various species lived. They are arranged so as to suggest a rough family tree, but the various lines of descent are controversial. For example, Brigitte Senut and her colleagues, who named *Orrorin tugensis*, have proposed that the line from *Ardipithecus ramidus* actually leads to the modern chimpanzee, while *Orrorin* leads through *Praeanthropus* to modern humans. They also suggest that the common ancestor of humans and chimpanzees dates from around 8.5 million years ago. These are indeed controversial claims, but nothing can be certain in so volatile a field of inquiry.

Apart from their bipedalism, the early hominins were probably not very different from their great ape predecessors. Increase in brain size and systematic tool cultures were to come later, with the emergence of the genus *Homo* a little over 2

[11] The discovery of Lucy's bones is described in Johanson and Edey 1981. She was named after the Beatles' song "Lucy in the Sky with Diamonds," whose initials are those of the hallucinatory drug LSD. It is to be hoped that Lucy was real, and not merely a hallucination.

[12] Richmond and Strait 2000.

[13] Leakey, Spoor, Brown, Gathogo, Kiarie, Leakey, and McDougall 2001.

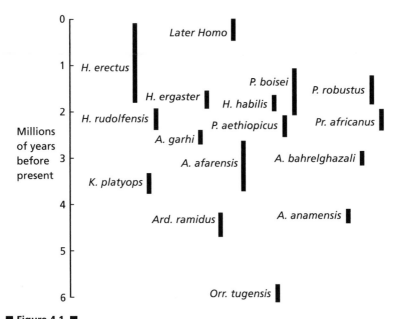

■ **Figure 4.1.** ■

Approximate dates of the various hominin species so far identified.
A. = *Australopithecus*; Ard. = *Ardipithecus*; H. = *Homo*; K. = *Kenyapithecus*;
Orr. = *Orrorin*; P. = *Paranthropus*; Pr. = *Praeanthropus*.

million years ago. Aside from revealing the remnants of a knuckle-walking phase, the fossil evidence also suggests that the hands and feet of the early hominins remained partially adapted to the grasping of tree branches; they may have still spent part of their time in trees, perhaps to escape dangerous predators. The full-striding gait of modern humans probably evolved only gradually, although even now the skill of circus performers on the trapeze reminds us that we still retain more than vestiges of our arboreal past.[14]

The fossil discoveries have amply confirmed Charles Darwin's hunch that we are descended from African apes. From their origin, perhaps 6 million years ago, until around 2 million years ago, the hominins were apparently confined to the African continent. Fossils dating from the end of that period have been found outside of Africa, first at Asian sites and later at European ones, but these evidently reflect a series of migrations out of Africa. Africa is truly the cradle of humanity.

[14] *Groves 1989 identifies four different stages in the evolution of bipedal walking.*

■ The savanna theory ■

No one knows for sure why the first hominins walked upright, although it must have had something to do with changes in the physical environment. Most archeologists believe that the principal change was from forest to more open land, known as *savanna*. In his 1925 article, Dart wrote that "Southern Africa, by providing a vast open country with occasional wooded belts and a relative scarcity of water, together with a fierce and bitter mammalian competition, furnished a laboratory such as was essential to this penultimate phase of human evolution." This so-called savanna theory has also been dubbed the East Side Story,[15] on the argument that the gradual opening of the Great Rift Valley in Africa effectively created different ecologies to the east and west. The hominins, according to this theory, were effectively trapped in the increasingly savannalike conditions on the east, while the other great apes continued to live in the wooded environments to the west.

By standing upright, our hominin ancestors may have been able to look out over the savanna for dangerous predators, such as hyenas or big saber-toothed cats. The lack of forested cover may also have meant that they had to travel longer distances over open terrain to forage for food, and bipedal walking is more efficient than knuckle walking. Nevertheless, one might ask why they didn't evolve a more efficient quadrupedal form of locomotion, like those of the other large mammals on the African plains, such as the fleet-footed antelopes. But apes are partly bipedal anyway, so the transition to full bipedalism may have been but a small step, albeit a giant step for mankind.

Another advantage of bipedalism is that the hands and arms can be used for carrying things. Foraging may have entailed carrying foodstuffs back to base, and an increasingly nomadic existence meant that these early hominins would also have had to carry their possessions, including their infants. Chimpanzee babies are less helpless than their human counterparts and can cling to their mother's bodies, whereas human babies must be cradled. Other species carry infants in pouches or in their mouths, but it would have required considerable change in anatomy to bring about such adaptations in the early hominins. All apes are fairly well adapted to holding things in their hands, and with the advent of bipedalism they would in effect have

---

[15] *Coppens 1994.*

been already prepared, or *preadapted*, to using their hands and arms for carrying.

But is the savanna theory really true? There seems to be growing support for an alternative theory, and with it, a different explanation for the emergence of bipedalism.

■ The savanna theory challenged: Water, water everywhere ■

In 1995 the eminent South African archeologist Philip V. Tobias surprised a large audience in London by declaring the "savanna hypothesis" to have been repudiated. He claimed that the fossilized plants found with remains of *Australopithecus* are not savanna vegetation. There has long been evidence that the early hominins actually lived in wooded environments, and that the shift to a more savannalike setting may not have begun until around 2 million years ago.[16]

Moreover, the evidence suggests that the australopithecines did not live in dry conditions, as Dart and others supposed, but inhabited forested areas that bordered on stretches of water. The 4.4-million-year-old hominin fossil *Ardipithecus ramidus* was discovered in the region known as the Middle Awash, on either side of the Awash river in Ethiopia,[17] and the nonhominin fossils associated with the next oldest, *Australopithecus anamensis*, suggest that the large proto-Omo river, which ran through the region, was bordered by forest.[18] Both *Orrorin tugensis*[19] and *Kenyanthropus platyops* also appear to have inhabited a lakeshore or river floodplain environment.[20] These early hominins may still have used trees for refuge and sleep but increasingly foraged in or near the water for food.[21] The later australopithecines of East Africa have also been associated almost exclusively with sites close to the shores of lakes or on floodplains or sandbars associated with rivers. Those in southern Africa are associated with caves, such as the Taung

[16] *Reviewing a book by Leakey and Harris 1987 on the paleoecology of Laetoli in Tanzania, where the remains of A. afarensis have been located, Andrews 1989, 180 remarks that "the habitat was more heavily wooded than any of the authors of this volume are prepared to acknowledge." WoldeGabriel, White, Suwa, Renne, Deheinzelin, Hart, and Heiken 1994 reviews the evidence on the habitat of A. afarensis around the Aramis riverbed in Ethiopia between 3.4 and 3.8 million years ago and concluded (p. 333) that "the Aramis hominins lived and died in a woodland setting." See also Rayner, Moon, and Masters 1993.*
[17] *White et al. 1994.*
[18] *Leakey et al. 1995.*
[19] *Senut et al. 2001; see also Balter 2001a.*
[20] *Leakey et al. 2001.*
[21] *See Tobias 1998. He actually goes so far as to claim that a paradigm shift has taken place, quoting Max Planck, the German physicist and Nobel laureate as saying: "A new scientific truth does not triumph by convincing its opponents and making them see the light, but rather because its opponents die, and a new generation grows that is familiar with it." In his 1995 address, the seventy-one-year-old Tobias went on to declare, "A change of paradigm shakes us up; it rejuvenates us; and, this above all, it prevents mental fossilization—and that is good for all of us." (continued)*

cave where *Australopithecus africanus* was found. These caves were formed as expansions of water channels that ran through the limestone.[22]

The idea that our ancestors went through an aquatic phase actually goes back over forty years. It was suggested by Sir Alister Hardy, a prominent British biologist, in 1960 and enthusiastically pursued by Elaine Morgan in several books (see Morgan 1990, 1997). The theory has not been widely accepted, perhaps because the claims for it have been exaggerated and in some instances based on inaccuracies. Among the human properties claimed as evidence for an aquatic phase are hairlessness, the presence of subcutaneous, voluntary breath control, a nose shaped so that one can dive without water going up it, and even bipedalism and language. For a detailed critique, see Langdon 1997. While offering encouragement to Morgan, Tobias actually suggests that the phrase "aquatic ape hypothesis" be dropped, since it immediately invites scorn. Tobias's proposal, like Hardy's original one, is much more cautious. Even so, I think Elaine Morgan deserves considerable credit for her persistence in the face of opposition from the biological establishment. She will probably turn out to have been more right than wrong.

[22] See Wood 1992 for further discussion of the australopithecine habitats. Although Wood does not specifically advocate the aquatic theory, he makes it clear that the australopithecines met watery graves.

[23] If Senut et al. 2001 are to be believed, however, this creature may have been a precursor of the modern chimpanzee, so the East Side Story is saved.

[24] Brunet, Beauvilain, Coppens, Heintz, Moutaye, and Pilbeam 1995. Ironically, it was Coppens, one of the authors, who had in the previous year argued in favor of the East Side Story; see Coppens, 1994.

Another reason to doubt the savanna theory was the discovery of a 3.5-million-year-old australopithecine fossil in Chad, which is some twenty-five hundred kilometers to the west of the Great Rift Valley—he was clearly a character from *West* Side Story.[23] Although closer in appearance to *A. afarensis* than to other hominins, he was sufficiently distinct to be named as a separate subspecies, now known as *Australopithecus bahrelghazali*. He was clearly bipedal, but he was found in a wooded area, raising doubts as to the role of the savanna in selecting for bipedalism. The discoverers of this fossil note that the remains of aquatic species were also discovered in the vicinity, and that the evidence generally is consistent with a lakeside environment.[24]

There are also some curious facts about modern chimpanzees that do not fit the savanna theory. Spurred by Raymond Dart's view that the australopithecines were shaped by a savanna habitat, the famous archeologist Louis Leakey initiated studies of modern great apes in the savannalike environments of southern and eastern Africa, hoping to gain further understanding of the hominin fossils discovered in those regions. Among those who took up the challenge were Jane Goodall, who studied chimpanzees, and Dian Fossey, who studied gorillas. These studies were at the forefront in the 1960s and 1970s, but more recently attention has turned to the behavior of chimpanzees and bonobos in the forested regions of west and central Africa. Ironically, it now appears that the chimpanzees of western Africa show more homininlike behaviors than the chimpanzees in the more savannalike environments. For instance, chimpanzees in the tropical rainforest of the Taï National Park on the Ivory Coast make and use more tools than those in the more open environments in the east. They use rocks and wooden hammers to crack nuts, producing artifacts that are very much like hominin artifacts. They are also more likely to modify their tools in advance, and they carry them longer distances. Selecting appropriate stone hammers and taking them to nut trees involves complex spatial knowledge—which is perhaps surprising, since it is generally assumed that spatial knowledge would be more critical to survival in the savanna than the forest. The Taï chimpanzees also hunt colobus monkeys more frequently and more cooperatively than the chimpanzees observed by Jane Goodall at Gombe in Tanzania in East Africa. All of these behaviors imply a higher degree of "hominization."[25]

If the early hominins inhabited forested land near the coast, or by lakes and rivers, then they may have turned to the water or the water's edge for sustenance, whether in the form of shellfish or marine plants. Bipedalism may then have been an adaptation to wading through water. By staying upright, an australopithecine could wade into deeper water than if forced to remain on all fours. Other primates also use a bipedal gait when wading. In fact, on land, the more common form of bipedalism is to hop like a kangaroo! Indeed, primates such as indris and tarsiers hop rather than walk, and if bipedalism had been an adaptation to locomotion over land, one might ask why the hominins didn't adopt that solution too. Kangaroos get about on land much more efficiently and rapidly than we do. But hopping is not an effective way to wade, unless you're a water bunny.

The idea that bipedalism was an adaptation brought about by wading, and perhaps swimming, seems to me to have much to recommend it. Other features of human anatomy that distinguish us from other primates, such as the absence of body hair, subcutaneous fat, long legs relative to body length, and a covered nose, likewise seem more consistent with an aquatic environment than a land-bound one. Bipedalism brings with it a number of woes, such as varicose veins, hemorrhoids, arthritic hips and knees, and spinal deterioration. Maybe we'd be better off if we went back to the beach.[26]

Yet not all of the evidence points to a wholly aquatic environment. The species *Australopithecus afarensis* is associated with three different sites, Laetoli in Tanza-

[25] *Boesch-Achermann and Boesch 1994.*

[26] *Where one should try to look one's best. As T. S. Eliot put it: "I shall wear white flannel trousers, and walk upon the beach."*

nia, Hadar in Ethiopia, and Bahr el Ghazal in Chad. The last two appear also to have been lakeside or riverside environments, but Laetoli does not seem to have been located near a water source.[27] Given the wide variety of locations at which fossil remains of these early hominins have been found, we might well conclude that they were rather restless, and versatile enough to move between different environments. Later, starting just under 2 million years ago, waves of hominins were to migrate from Africa and inhabit even more diverse environments, and the ability to adapt to different conditions may be our true heritage.

All this leaves us still somewhat in the dark, or perhaps treading water, as to why bipedalism evolved in what is emerging as a diverse family of creatures. But however it came about, our upright stance ensured that our destiny was more than ever before in our own hands. Those versatile, flexible appendages, already subject to a high degree of voluntary control through tens of millions of years of adaptation to a forested environment, were released for other duties. One skill that may have emerged as a consequence of bipedalism is the ability to throw missiles, with potentially lethal accuracy.

■ Throwing ■

Throwing is certainly an important human activity, whether in recreation or in anger. It is not, however, an exclusively human activity, and other primates are reasonably good at it. For example capuchins, a species of New World monkey found in South and Central America, can throw stones at both moving and stationary targets with considerable accuracy and can use both power and precision grips for throwing. They also use throwing as a way of transferring food between social groups, much as present-day Australians do at a barbecue.[28] In one study,[29] capuchins proved quite good at throwing stones into a bucket partially filled with either peanut butter or a sweet syrup, and the reward for accuracy was being allowed to lick the stone after it landed in the goo. About half the time they threw from an upright, bipedal position, and most of the

---

[27] See Leakey et al. 2001 for more detailed speculations about the habitats in these four regions. It is also worth noting that the fossils of the newly claimed hominin Orrorin tugensis (see footnote 5) were discovered in lake and river sediments, again suggestive of an aquatic environment (Balter 2001a).

[28] As an undergraduate, I spent some time in a dormitory loosely modeled on an Oxbridge college. I can recall the master, imported from Cambridge to instill gentlemanly virtues in unruly colonials, vainly appealing to the mob in the dining hall: "Gentlemen will kindly desist from throwing the food."

[29] Westergaard, Liv, Haynie, and Suomi 2000.

time they threw overarm. Even so, they weren't as proficient as the humans who were tested, and a number of other interesting differences came to light as well. Although all the capuchins threw with one hand, the group as a whole contained as many left-handers as right-handers, whereas humans are overwhelmingly right-handed—with of course some conspicuous exceptions in sports such as baseball, cricket, tennis, and football. Female capuchins were just as proficient as males, whereas in our own species males excel even in the three- to five-year-old age group, suggesting that in humans the sex difference in throwing ability is at least partly biological rather than cultural.[30]

Throwing has not been studied quite so extensively in chimpanzees, but they can certainly hurl objects, such as branches of trees, as a means of self-defense. Nevertheless Charles Darwin writes, "As I have repeatedly seen, a chimpanzee will throw any object at hand at a person who offends him."[31] It is interesting to observe that, although the bonobo Kanzi was taught to make stone tools in Oldowan fashion, he actually seems to prefer to make stone flakes by flinging the stone against a hard surface, as capuchins sometimes do.[32] Again, though, the way chimps and bonobos throw is crude compared with the way humans do and is perhaps better described as a fling than a throw. It has none of the precision and power evident among the flanneled fools who indulge in the sober but surprisingly dangerous game of cricket. In baseball, a pitcher can throw a fastball at ninety miles per hour, and a batter can (sometimes) hit a fastball on the fly with a rather narrow bat. These are remarkable skills, and must surely have evolved for reasons other than the amusement of television audiences.

Mary Marzke has argued that changes in bodily structure and posture for the enhancement of accurate throwing can be traced as far back as *Praeanthropus africanus* (previously known as *Australopithecus afarensis*) over 3 million years ago.[33] Not only did the structure of the hand change in ways consistent with holding and hurling rocks or other fist-sized objects, like modern-day baseballs or cricket balls, but the bipedal posture would have supplied extra leverage. Although baseball and cricket are rather recent inventions, it is conceivable that the bipedal stance itself was, at least in part, the result

[30] See Kimura 1999 for this and other evidence.
[31] Darwin 1896, 82. Darwin also mentions a baboon at the Cape of Good Hope who not only threw missiles at people but prepared mud for the purpose.
[32] Toth, Schick, Savage-Rumbaugh, Sevcik, and Rumbaugh 1993.
[33] Marzke 1996.

of selective pressure for more effective throwing and clubbing. Certain details of the structure of the human leg may also have more to do with throwing than with walking or running. Our legs are relatively more massive than those of the ostrich, say, and the knee includes a locking device that presumably has little to do with locomotion. These adaptations may provide the stable launching platform needed for strong and accurate throwing.[34]

If the early hominins were indeed forced to adapt to the savanna, they would have been easy (and no doubt delectable) meat for the saber-toothed cats and hyenas that patrolled the region, and bipedal throwing may have been critical to survival. Our primate ancestors, remember, were adapted to climbing trees, but on the open savanna would have had less opportunity to escape by the traditional method of shinning up a tree. Even so, trees still sometimes provide a useful means of escape, as illustrated by a famous incident in South Africa in 1903. Harry Wolhuter was a game ranger in the Kruger National Park, and while riding on horseback was set upon by two lions. One lion attacked his horse, throwing Harry to the ground, and the other lion then grabbed him by the right shoulder and started dragging him off. In the ensuing struggle, and after being dragged sixty yards, he managed to kill the lion with a knife and then climb a nearby tree, where he strapped himself to a branch. This enabled him to escape an attack by the other lion, with some assistance from his dog, who worried the lion as it tried to reach him. Eventually help arrived and Harry lived to tell the tale—and many others, I'm sure, since he remained alive until 1964 and died just before his eighty-eighth birthday.

The dangerous environment of the savanna, perhaps still the stuff nightmares are made of, may have created pressure to ward off predators by throwing rocks or branches, initially in the crude fashion of present-day chimpanzees, but with increasing skill and precision. Even so, it is doubtful whether Harry Wolhuter would have been able to ward off the lions by throwing rocks at them, and the argument loses some of its force if the early hominins actually remained in a forested environment near water, as Tobias and others now argue. If this were the case, they might still have been able to escape predators by climbing trees, as Harry did, or by retreating into the water. Throwing, then, might have had a more benign aim, so to speak. For example, our resourceful ancestors may have collected shellfish and

---

[34] *Fifer 1987.*

thrown them to the shore, where they could be collected later, or picked up by their cronies.

But throwing undoubtedly became aggressive at some point. Again, perhaps, the critical change may have come 2 to 3 million years ago, when survival began to depend more on foraging away from forested or waterside environments, perhaps to pick over carcasses left from a kill of an antelope by lions, say. In such a situation, rocks may have been thrown to keep other predators at bay. Eventually, though, an increasing ability to throw may have turned defense into attack. Marzke even suggests that early stone tools, dating from about 2.5 million years ago, may have been used not only as choppers and scrapers but also as objects to be thrown for the kill. Spears are, of course, designed as killer weapons and go back at least 400,000 years in human history.[35] Even in modern times there is a temptation to throw things at those who annoy. It is just as well, for example, that William Shakespeare did not live to be a contemporary of the vastly more talented George Bernard Shaw, for he would have had to duck more than just the slings and arrows of outrageous fortune: "With the single exception of Homer, there is no eminent writer, not even Sir Walter Scott, whom I can despise so entirely as I despise Shakespeare when I measure my mind against his. . . . It would positively be a relief to me to dig him up and throw stones at him."

People are indeed quick to resort to throwing things to express aggression. The television news often features angry crowds in the trouble spots of the world throwing stones, rocks, or bottles at those they love to hate. Nevertheless, in modern industrial society, those of us who are not professional players of ball sports may have lost some of the throwing ability of our forebears. There is evidence that people in less complex societies, who practice from an early age, can throw projectiles with an accuracy and speed that would astonish modern-day couch potatoes.[36] According to the eighteenth-century explorer J. W. Vogel, the Hottentots of southwestern Africa "know how to throw very accurately with stones. . . . It is also not rare for them to hit a target the size of a coin with a stone at 100 paces."[37] Australian aboriginals were also said to be able to throw stones with enough accuracy and force to bring down wallabies and flying birds, dislodge nuts from the baobab tree, and knock fledging birds out of high nests.

Paul Bingham has argued that the ability to throw accurately gave

[35] *Thieme 1997.*
[36] *Isaac 1987.*
[37] *Quoted in Isaac 1992.*

our forebears the unique ability to kill at a distance, with profound consequences for our family tree. But that story must wait for the following chapter, where we will consider what happened in hominin evolution over the past 2 million years.

■ Bipedalism and language ■

William H. Calvin, in his book *The Throwing Madonna*, has suggested that throwing may have set the stage for language.[38] Like speech, throwing requires very accurate timing, with precise adjustments for direction and distance. Most people can throw with only one arm, usually the right, and the development of throwing skill would have led to the emergence of appropriate circuits for timing in the opposite cerebral hemisphere. Calvin suggests that this may be why speech is also represented in only one side of the brain, usually the left, in the great majority of people. I have doubts as to the generality of this theory,[39] but given that throwing is a manual gesture, this theory would surely support a closer link between throwing and gestural language than between throwing and speech. Save for the hurling of insults, people do not throw well with their mouths.

But quite apart from any relation between throwing and gesture, bipedalism itself would surely have enhanced gestural communication.[40] Indeed, it may well be that the early hominins developed a gestural protolanguage in the 2 million or so years following the split from the chimpanzee line. We have already seen that present-day chimpanzees and bonobos are at least capable of protolanguage, although there is little evidence that they use it spontaneously in the wild. The emergence of bipedalism may have given us that extra nudge, so to speak, that set us on the path to language. It is unlikely, though, that the hominins evolved true grammatical language prior to the emergence of the genus *Homo* some 2 million years ago.

The early hominins bequeathed to us a bipedal posture, and perhaps some skill at throwing and gestural expression, but otherwise they

---

[38] *Calvin 1983.*

[39] *Calvin repeats the suggestion in a new book (Calvin and Bickerton 2000), but his coauthor, Bickerton, objects that throwing does not have the open-ended, generative property of language. Calvin's earlier book,* The Throwing Madonna, *also implies that throwing evolved among women, largely as a means of warding off would be attackers. As we have seen, however, males are better than females at throwing, even in early childhood, which suggests that throwing probably evolved in the context of hunting. But women are undoubtedly better talkers.*

[40] *In one respect it might also have contributed to the evolution of articulate speech. By altering the way the skull is positioned on the spinal column, upright posture may have been indirectly responsible for the lowering of the larynx, which greatly increased the range of articulate sounds we can produce. This point is elaborated in chapter 7.*

remained ape like. The critical event that ended the early era may have been a global shift to cooler climates some 2.5 million years ago, which transformed wooded and possibly watery habitats to more open, grassy ones. The more robust varieties of early hominin were adapted to a vegetarian diet, with heavy jaws and teeth for grinding hard roots. Indeed, two of our sturdy cousins, *Paranthropus robustus* in southern Africa and *Paranthropus boisei* in eastern Africa, appear to have survived in the more savannalike environment until a little over a million years ago.[41] Among the so-called gracile hominins, adapted more to eating fruit and perhaps shellfish, more dramatic changes took place, leading eventually to the evolution of modern humans. These changes included bigger brains, the development of tool technologies, migrations from Africa, and more "advanced" thought processes—and, of course, language.

These are the topics of the next chapter.

[41] *Wood 1992.*

Perfectibility is one of the most unequivocal characteristics of the human species.
—William Godwin, *An Enquiry Concerning the Principles of Political Justice* (1793)

5 ■ Becoming Human ■

Although bipedal hominins emerged some 5 or 6 million years ago in Africa, we have relatively little evidence that they evolved anything resembling modern human behavior for the next 3 or 4 million years. Bipedalism may have allowed them to be more expressive in their gestural communication, but they probably remained essentially chimpanzeelike in other respects. There is no compelling reason to suppose that their communication evolved beyond protolanguage: the ability to form representations and combine them in short sequences, but without the grammatical complexities that distinguish human language. It should perhaps be remembered that, at the molecular level, even modern humans bear a closer resemblance to chimpanzees than chimpanzees do to gorillas, leading Jared Diamond to refer to us as the "third chimpanzee."[1]

Important changes began to manifest some 2 million years ago, with the emergence of the genus *Homo*, and these may signal the first advances toward a more sophisticated grammatical language. These changes are the topic of this chapter. Somewhat in limbo, though, are the earliest members of the genus *Homo*, namely *Homo rudolfensis*, dating from about 2.5 million years ago, and the slightly later *Homo habilis*. It has been argued that these two do not really belong to the genus *Homo* at all and should be reclassified as australopithecines.[2]

[1] *J. Diamond 1992. The other is the bonobo.*
[2] *Wood and Collard 1999.*

That is, the true climb—or descent, if you prefer—to humanity may be said to have begun with two other species, *Homo ergaster* and *Homo erectus*, both dating from around 1.9 millions years ago. *Homo ergaster* was once considered to be identical to early African *Homo erectus* but is now more commonly considered a separate species, which existed until around 1.5 million years ago.[3] *Homo erectus* appears to have been more peripatetic, migrating to Asia perhaps as early as 1.9 million years ago, and was certainly more durable, as we shall see below.

What, then, were the humanlike characteristics that began to emerge with the genus *Homo* around 2 million years ago? It was probably not until some time in this final stage that true language began to emerge: we walked the walk well before we talked the talk. But researchers still cannot agree as to precisely when grammatical language emerged. Whether vocal or gestural, language leaves little trace in the archeological record, and we must look to other evidence for what happened as our species took the next step toward humanity. Moreover, as we have seen, we can learn relatively little about true language from the communicative skills of other species, as even our closest relatives, the chimpanzee and bonobo, do not appear to be capable of more than a crude form of protolanguage.

One characteristic of humanness that has left more tangible traces is the ability to make tools. This is an activity that may well have further enhanced the gestural repertoire, although, as we shall see, it may not have been quite as dramatic a development as was once believed.

■ Toolmaking ■

We are not, in fact, the only species that makes tools. Indeed, some of the most proficient tool makers are not primates but—you've guessed it—birds. New Caledonian crows show an extraordinary capacity to cut leaves from the pandanus tree and fashion them into hooks for extracting grubs from cracks and holes in trees. These tools appear to be deliberately given a stepped, tapered shape, with one end wide enough to be held in the beak and the other more pointed, so it can be inserted in the hole or crack. The leaves have a saw-tooth edge, and the tool is cut out so that the teeth point back from the narrow end and can therefore attach to the prey and allow it to be pulled out. The crows make several different tools of this kind, apparently constructed according to premeditated designs. Most of the tools are cut out from the left edge rather than the right edge of the leaf. This implies that the bird favors the right eye in guiding its handiwork, which suggests that the left side of the brain is specialized for this task.[4]

Well, stone the crows, I say.[5] No wonder Robert Greene, the Elizabethan hack writer known for his dislike of Shakespeare, referred to the bard as "an upstart crow, beautified with our feathers."[6] And perhaps Shakespeare had Greene in mind when he wrote, "If a crow help us, sirrah, we'll pluck a crow together."[7]

But it's not just the crows. Animals have been widely observed to use natural objects as tools. Charles Darwin wrote of tamed elephants in India who broke off the branches of a tree and used them to drive away flies. Ever the naturalist, he once observed "a young orang put a stick into a crevice, slip his hand to the other end, and use it in the proper manner as a lever."[8] Capuchin monkeys, mentioned in the previous chapter for their ability to throw, have been seen to use tools in the wild without any human intervention, and in novel and isolated ways that suggest improvisation rather than social training. These include such acts as using a stick to kill a snake and rocks to crack open oyster shells.

More tellingly, perhaps, capuchins have also been observed to *make* probes out of sticks and to modify stones and bone fragments for cutting or for cracking nuts. They make stone flakes by striking stones against each or against other hard surfaces, or by throwing stones from a high perch onto a hard floor. These observations have led some to argue that toolmaking is not simply a product of the progression from monkey to ape to hominin but could only have emerged in omnivores.[9] Other evidence suggests that capuchins are incapable of transporting tools to food sites, perhaps because they lack the foresight and capacity for mental representation shown by the early hominins.[10] Chimpanzees also demonstrate a variety of tool techniques, such as the use of twigs as probes for termites or ants, leaves as sponges, and rocks and pieces of wood as hammers to crack nuts.[11] We saw in the previous chapter that the best toolmak-

[4] These observations were made by my intrepid colleague Gavin Hunt 2000.
[5] An Australian expression, I believe.
[6] From his play Groats-Worth of Witte 1592.
[7] From Comedy of Errors, act 3, scene 1. The expression "I have a crow to pluck with you" meant the same as "I have a bone to pick with you." And no doubt this is how Will felt about Robert.
[8] Darwin 1896, 83.
[9] Westergaard 1998.
[10] Jalles-Filho, Da Cunha, and Salm 2001.
[11] For review, see Tomasello and Call 1997.

ers among the chimpanzees are those inhabiting the wooded environment of West Africa; this would suggest that the savanna was not especially critical to the emergence of toolmaking.

It is unlikely that the early hominins were much more sophisticated than present-day chimpanzees in using or making tools, although the upright stance may have given them something of a helping hand, so to speak. If they foraged in or by water, as Tobias has claimed, then they might have used rocks to crack open shellfish, as capuchin monkeys do today.[12] But if they fashioned tools out of perishable materials, such as wood or leaves, there would be no traces in the fossil record. Although they probably did make and use tools, my guess is that tools and toolmaking were not especially critical to the evolution of the early hominins. They were more like great apes than like humans and had brains similar to the chimpanzee's.

The first clear sign of advance was the emergence of stone tools that were evidently fashioned for specific purposes and made to last. The fact that stone tools *are* part of the fossil record may have created a false impression of their importance, since it is conceivable that equally sophisticated but perishable tools had been manufactured earlier. Nevertheless, stone tools are associated with the genus *Homo* and not with the other hominin genera, such as the australopithecines, and in other ways *Homo* seems to represent a significant departure from apehood and toward humanity, as we shall see below. It is also possible that the development of stone tools reflects a decreasing reliance on water-based resources, perhaps due to the retreat of forested areas and a growing scarcity of marine-based foodstuffs. That is, the happy life on the beach, or by the river, may have been drawing to a close 2 or 3 million years ago.

The first known stone tool industry consisted of simple stone flakes, found initially at Olduvai Gorge in Tanzania, and known as the Oldowan industry. The earliest tools in the Oldowan style have been discovered in Gona in Ethiopia, not Tanzania, and dated at between 2.5 and 2.6 million years ago.[13] Oldowan tools were at first associated with *Homo habilis* ("handy man"), but the Ethiopian collection predates this species and may be associated with *Homo rudolfensis*, the species presently considered to be the first of the *Homo* line—although, as we saw above, there is some doubt as to whether we should really include *rudolfensis* in the genus *Homo*, and perhaps, like Groucho Marx, he wouldn't want to join the

[12] *Fernandes 1991.*

[13] *Semaw, Renne, Harris, Feibel, Bernor, Fesseha, and Mowbrau 1997.*

club anyway. Oldowan tools were probably used for cutting and scraping carcasses of animals that had been killed by other predators.

But the Oldowan industry itself probably does not represent a major intellectual advance. As we saw above, capuchins have also been observed to make flake tools. Moreover the bonobo Kanzi, whom you were privileged to meet in chapter 2, has recently learned to make stone tools equivalent to those of the Oldowan industry.[14] Kanzi was shown how to do this by his human guardians, however, and there is no evidence for the spontaneous manufacture of stone tools by any of the great apes (except for humans). About 1.5 to 1.7 million years ago, a more sophisticated industry, the Acheulian, developed in Africa. This industry is associated with *Homo erectus*. Acheulian artifacts include large cutting tools, picks, cleavers, and bifacial hand axes, quite possibly hafted. An analysis of plant residues on the surfaces of hand axes found in Tanzania indicates that they were used for woodworking.[15]

It has also been suggested that the Acheulian industry signals the emergence of hunting, and that early *Homo*, armed with more sophisticated tools and weapons, migrated out of Africa in search of game. However, we saw earlier that some groups of *Homo erectus* began to migrate out of Africa well before the Acheulian industry developed, perhaps as early as 1.9 million years ago. There is controversial evidence that *erectus* reached Java as early as 1.8 million years ago[16] and may have survived there as a species until as recently as 27,000 years ago.[17] If these dates are correct, they make *Homo erectus* the most successful so far of the genus *Homo*, and we shall be lucky if we last as long. Fossils comparable to those of *Homo ergaster*, the early African cousin of *erectus*, have been unearthed in the Republic of Georgia, and recently dated at about 1.7 million years.[18] The tools associated with these remains are Oldowan rather than Acheulian, consisting of simple scrapers and choppers. There were probably several waves of migration out of Africa,[19] with at least one wave going east and others heading eventually toward Europe (figure 5.1). Since there is no evidence that the tool industry of the early migrants, at

[14] Toth, Schick, Savage-Rumbaugh, Sevcik, and Rumbaugh 1993.

[15] Dominguez-Rodrigo, Serrallonga, Juan-Tresserras, Alcala, and Luque 2001.

[16] Swisher, Curtis, Jacob, Getty, Suprojo, and Widiasmoro 1994.

[17] Swisher, Rink, Anton, Schwarcz, Curtis, Suprijo, and Widiasmoro 1994. If this date is correct, they would have overlapped with Homo sapiens in the region. The dismal conclusion is that H. sapiens probably exterminated them. Of course, they may have done so unwittingly, perhaps by introducing diseases that the incumbents had no defenses against.

[18] Gabunia, Vekua, Lordkipanidze, Swisher, Ferring, Justus, Nioradze, Tvalchrelidze, Anton, Bosinski, Joris, de Lumley, Majsuradze, and Mouskhelishvili 2000.

[19] Tattersall 1997.

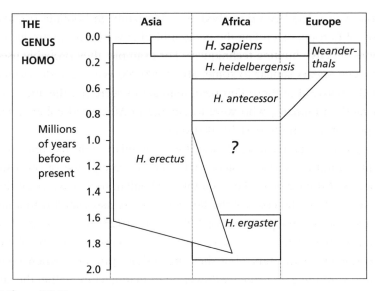

THE GENUS HOMO

	Asia	Africa	Europe
0.0		H. sapiens	Neander-thals
0.2		H. heidelbergensis	
0.6		H. antecessor	
0.8			
Millions of years before present 1.0	H. erectus	?	
1.2			
1.4			
1.6		H. ergaster	
1.8			
2.0			

■ **Figure 5.1.** ■

Emergence and spread of various species of *Homo*.

least, had developed beyond the Oldowan, these initial migrations may have been driven by foraging, perhaps along shorelines, rather than by scavenging carcasses or by hunting.

Some Acheulian assemblages are found outside of Africa, however, notably in Israel, which was presumably on a later migration corridor from northeastern Africa to southwestern Asia. The Israeli site at Ubeidiya has been dated at 1.4 million years, the one at Evron Quarry at about 1 million years, and the one at Gsher Benot Ya'aqov in the Dead Sea Rift at 780,000 years. The last of these assemblages was found following the draining of the Hula Lake and thus indicates a lakeside habitat. Remnants of one hundred varieties of seeds and fruits, many of them derived from water-based plants, were found at the site, along with the hand axes and cleavers characteristic of Acheulian tools in Africa.[20] Very recently, evidence has been discovered of an Acheulian tool culture associated with *Homo erectus* in South China.[21]

A relatively sophisticated Acheulian industry is associated with the appearance of in Europe, around 600,000 years ago, of *Homo heidelbergensis*. The arrival of this species, probably originally

[20] Goren-Inbar, Feibel, Verosub, Melamed, Kislev, Tchernov, and Saragusti 2000.

[21] Until recently, it was thought that the Acheulian industry was restricted to western Eurasia and Africa, but a recent study by Yamei, Potts, Baoyin, Zhengtang, Deino, Wei, Clark, Guanmao, and Weiwen 2000 provides rare evidence for Acheulian technology in East Asia.

from Africa, has been described as a "big bang" in hominin occupation of Europe, extending the geographical coverage and introducing more effective hunting techniques.[22] For example, they probably used wooden spears similar to some, about 400,000 years old, discovered in Germany.[23] But these European immigrants were not the ancestors of modern humans, who were to emerge in Africa and colonize Europe only within the past 100,000 years.

Meanwhile, back in Africa, the Acheulian industry remained quite primitive, and persisted until after the emergence of our own species, *Homo sapiens*. For example, Acheulian tools have been discovered along the coast of the Red Sea and are associated with a migration of *Homo sapiens* out of Africa around 125,000 years ago.[24] Of course, these stone tools may not give a complete picture of the technology of the time, but they are indeed primitive when compared with the extraordinary flowering of manufacture within the last 50,000 years.

The relation between language and tools has long been a matter of speculation. Some have argued that manufactured tools signal the prior emergence of language, or at least protolanguage.[25] Merlin Donald has argued, however, that this puts the cart before the horse, and that language may have emerged rather from the programming of sequences involved in such activities as tool use and throwing.[26] But this too seems unlikely, since there is little apparent correlation between the sophistication of language and that of tool manufacture, even today. Tools are more likely to be a product of environmental challenges than of language ability per se. My own view is that a more sophisticated language probably did emerge over the past 2 million years and was accompanied by more innovative and generative thought processes that enabled the development of more sophisticated tools when circumstances demanded it. However, language was probably primarily gestural rather than vocal, so the development of manufacture may have actually been *inhibited* by the use of the hands for communication. This could be why manufacture did not really begin to flower until speech became the dominant mode, perhaps as

[22] See Balter 2001b.
[23] Thieme 1997. These spears were probably manufactured by Neanderthals, or perhaps H. heidelbergensis.
[24] Walter, Buffler, Bruggemann, Guillaume, Berhe, Negassi, Libsekal, Cheng, Edwards, von Cosel, Neraudeau, and Gagnon 2000.
[25] Holloway 1969 suggested that the elements of design inherent in stone tools, more especially in the Acheulian tools of Homo ergaster than in those of the more primitive Oldowan culture, imply a dependence on language. Similarly Dennett 1992 has suggested that these early hominins could only have made a complex tool by "talking to themselves" in some form of language, perhaps protolanguage.
[26] Donald 1999.

■ Table 5.1 ■
Estimates of average brain size of various great apes and hominins

Species	Body weight (kg)	Brain volume (cc)
Human	67.7	1,355
Neanderthal	76.0	1,512
Homo heidelbergensis	62.0	1,198
Homo erectus	57.0	1,016
Homo ergaster	58.0	854
Homo habilis	34.0	552
Homo rudolfensis	unknown	752
Chimpanzee	55.4	337
Bonobo	45.4	311
Gorilla	61.7	397
Orangutan	73.5	407

recently as 50,000 years ago, thereby freeing the hands. But that is a story for later, in chapter 9.

■ Bigger brains ■

The emergence of the genus *Homo* is important not only because it coincides with the earliest known stone tools but also because it marks the beginning of an increase in brain size. The australopithecines and other early hominins had brains that were about the same size as those of the great apes, at least when body size is taken into account, but the genus *Homo* became increasingly swellheaded, as table 5.1 shows.[27]

Our brains are over three times as big as one would expect of an ape with the same body size.[28] Indeed, this is perhaps the most dramatic difference between ourselves and our nearest living cousins: we are truly brainier. Whether wiser I don't know. Curiously, the Neanderthals, who became extinct some 30,000 years ago, had slightly larger brains than do we modern humans, although the difference effectively disappears if it is corrected for the fact that their bodies were also slightly larger.

The increase in brain size might actually owe something to an aquatic environment. The development of the brain depends on the accumulation of a complex fatty acid called docosahexaenoic acid (DHA for

[27] The figures for the great apes are taken from a study by Rilling and Insel 1999 and rounded slightly, while those for the various members of the genus Homo are taken from Wood and Collard 1999.

[28] Passingham 1982.

short). This is normally synthesized internally in the course of development, but it has been claimed that human infants cannot synthesize enough DHA for their own brain development unless they receive it from external sources. DHA is in short supply in an inland food chain but is readily available from shore-based foods or fish. If your mother told you fish was good for your brain, she was right, although by now it may be too late. The plentiful supply of DHA in the marine environment inhabited by the hominins of 2 or 3 million years ago may therefore have been necessary for the evolution of our large brains.[29]

But DHA was probably not a *sufficient* reason for the increase in brain size. The selection of larger brains was no doubt favored by other eventualities, perhaps related to shifts in the ecology, as I shall suggest below. Another change that might have occurred around this time was an increase in the duration of childhood, allowing a longer period of growth outside the womb. Relative to chimpanzees and other primates, humans are born prematurely. Indeed, to conform to the general primate pattern, human babies should be born at around eighteen months, not nine months, but as any mother will know, this would be impossible, given the size of the birth canal. Not for nothing is the process of giving birth called labor. The brain of a newborn chimpanzee is about 60 percent of its adult weight, that of a newborn human only about 24 percent. Our prolonged childhood means that the human brain undergoes most of its growth while exposed to external influences and is therefore more finely tuned to its environment. Moreover, this allows the brain to grow larger relative to body size than it does in other primates. The evidence suggests that this lengthening of childhood was present in *Homo erectus* by about 1.6 million years ago,[30] and also in the Neanderthals, who lived until about 30,000 years ago, but not in *Homo habilis* or *Homo rudolfensis*.[31]

The increase in brain size and the prolongation of childhood seem much more obviously related to mental development, and especially language, than to tool technology. As we have seen, tools seem to have remained quite primitive even among early *Homo sapiens*. In most people, much of the left cerebral cortex seems to be devoted to language in one way or another, suggesting that language does require a fair amount of brain space—although evolution has cunningly arranged to conserve brain space by confining

[29] Broadhurst, Cunnane, and Crawford 1998.
[30] Brown, Harris, Leakey, and Walker 1985.
[31] Wood and Collard 1999.

language mechanisms largely to a single hemisphere. Moreover, as I suggested in chapter 1, the development of recursive grammar may depend on an interaction between learning and brain growth, made possible through the prolongation of childhood.

■ The changing scene ■

A little over 2 million years ago, then, a number of changes began to occur. Stone tools appear in the fossil record, the brain increased in size, and hominins migrated from Africa. It is likely that grammar was invented. Why did things begin to change at this point in hominin evolution? Bipedalism alone was probably not a critical factor. After all, the early hominins had been bipedal for over 3 million years, without obvious signs of change in brain size or the disposition to manufacture stone tools.

It was suggested in the previous chapter that the early hominins inhabited wooded environments near water and foraged for water-based food. With the global shift to cooler climates after 2.5 million years ago, Africa became more open and sparsely wooded.[32] Some hominins may have been able to maintain a semiaquatic lifestyle by migrating northward round the coasts and continuing to forage for marine-based foodstuffs. There are good reasons to suppose that they traveled at least some of the way by water, perhaps strolling, wading, and even swimming. Fossils and stone tools on the Indonesian island of Flores indicate that *Homo erectus* had arrived there 800,000 to 900,000 years ago.[33] Even though the sea level was lower then than now, *erectus* would still have had to cross a deep oceanic channel some nineteen kilometers wide, perhaps by floating on logs or, again, even by swimming. Some stone tools and hominin remains in southeastern Spain date from over a million years ago, and it has been speculated that the hominins traveled there by crossing the Strait of Gibraltar, which would then have been about five kilometers across.[34]

It is generally agreed, however, that these migrants were not the forebears of modern humans. The so-called "Out of Africa" scenario, first proposed by Christopher Stringer and Peter Andrews, is that *Homo sapiens* evolved in Africa and later migrated, eventually spreading to all parts of the world, replacing *Homo erectus* in Asia and the

[32] *Wood 1992.*
[33] *Morwood, O'Sullivan, Aziz, and Raza 1998.*
[34] *Tobias 1998.*

Neanderthals in Europe.[35] Rather than joining the earlier exodus, then, our forebears remained in Africa, probably until as recently as about 52,000 years ago[36] and adapted to savannalike conditions there. The period during which this adaptation occurred overlaps largely with the epoch known as the Pleistocene, dating from about 1.8 million years ago until 10,000 years ago.[37] Evolutionary psychologists, such as John Tooby, Leda Cosmides, and Steven Pinker, have proposed that it was during the Pleistocene that the main characteristics of the human mind evolved, as our forebears adapted to a hunter-gatherer way of life.[38] These hominins would have been relatively ill-equipped physically for life on the savanna, so they evolved cognitive strategies for survival. Their place on the savanna came to be what evolutionary psychologists have termed the "cognitive niche": they lived by their wits rather than their muscles.[39] They would have had to cope with the dangerous killers, such as saber-toothed cats and hyenas, that roamed the plains of eastern and southern Africa, and they may have at first scavenged and then hunted for meat. This could be why the Acheulian culture was more prevalent in Africa than among the more leisurely beach strollers, who may have simply kept on going round the coasts when food ran out in any one place. The more challenging environment of Africa may partly explain why that all-conquering species, *Homo sapiens*, when they eventually did leave Africa, were able to adapt to their newfound territories and displace those hominins who had migrated earlier.

For other species on the savanna, avoiding predators usually depends simply on keen senses, to detect their presence, and fleetness of foot, to escape. Further, most predators hunt only when they are hungry and rely on physical cues to detect and track down their prey. The hominins, by contrast, evolved more cognitive strategies for survival. They planned their predatory activities in advance to minimize danger and maximize attacking opportunity. Tom Suddendorf has remarked that "while a full-bellied lion is no threat to nearby zebras, a full-bellied human may well be."[40] Per-

---

[35] Stringer and Andrews 1988. The out-of-Africa scenario has nevertheless been intermittently challenged, most notably by Milford Wolpoff, who has argued for a so-called "regional-continuity" scenario, in which modern humans evolved from earlier migrants, including *Homo erectus*, in different parts of the world. For his most recent defense of this model, see Wolpoff, Hawks, Frayer, and Hunley 2001.

[36] Ingman, Kaessmann, Pääbo, and Gyllensten 2000.

[37] Jones, Martin, and Pilbeam 1992. These authors also note that the Pleistocene is still thought by some to date from about 1.8 million years ago.

[38] Tooby and Cosmides 1989. See also Pinker 1997 for a popular account of evolutionary psychology.

[39] Okay, you want a better definition? Tooby and DeVore 1987 define the cognitive niche as a capacity for "conceptually abstracting from a situation a model of what manipulations are necessary to achieve proximate goals that correlate with fitness" (p. 209). Sorry you asked?

[40] Suddendorf 1999, 242.

haps this is why we humans are relentlessly preoccupied with time, going to the supermarket when time permits and not just when we're hungry, while other creatures just seem to laze about a lot of the time. Especially cats.

■ Cooperation ■

Perhaps the most important survival technique developed by our forebears was an extensive ability to cooperate, which can have extraordinary advantages in terms of fitness, general well-being, and the overcoming of physical odds. I suspect most of us would not survive were it not for the cooperative endeavors of our fellow humans, ensuring that we have food, clothing, housing, ways of preventing or curing disease, and the Internet. Large-scale cooperation does appear to be unique to humans, at least among large-bodied animals, and may well be the source of our distinctive mental abilities, including language.

Animals other than humans do occasionally show cooperative behavior, but it is much more restricted than it is in humans. An animal may come to the aid of another animal in a fight, forming a coalition to defeat an attacker, but such coalitions are usually between close genetic relatives. Such behavior is *altruistic*, since the helper may risk injury or death, but it can nevertheless sometimes be explained in genetic terms. The late William D. Hamilton pointed out that a gene that induces altruistic behavior, risking death or at least loss of progeny for its possessor, may nevertheless spread through a population if the altruistic behavior helps a relative carrying the same gene to produce more offspring.[41] In short, in both humans and other animals, coalitions between kin can be sustained by genetic mechanisms. This can explain why we put ourselves out for our own children and other close relatives, but it does not explain why we are often altruistic toward unrelated individuals.

Nonhuman animals also occasionally form coalitions that do not involve close relatives, although these coalitions tend to be unstable. For example, unrelated male baboons may form coali-

[41] *Hamilton 1964. More precisely, Hamilton's rule states that an altruistic gene that lowers the fitness of its possessor will spread in the population if $rB > C$, where B is the benefit, or the number of offspring made possible by the altruistic act, r is the degree of relationship between the individual performing the act and the beneficiary, and C is the cost, or the number of offspring the altruist loses by performing the act. J.B.S. Haldane put it more simply: when asked whether he would lay down his life for another, he replied that he would not, but that he would do so for two of his brothers or four of his cousins.*

tions for protection against more powerful males, or a low-ranking male chimpanzee may form a coalition with a second-ranking male to overthrow the top-ranking male.[42] Such coalitions may represent something called *reciprocal altruism*: the participant's understand that their altruistic behavior will be rewarded on some future occasion, on the principle of "you scratch my back and I'll scratch yours." But there is always the risk of deception or shifting alliances, and coalitions between nonkin readily come apart. In human society, we have a number of terms to describe the various ways in which coalitions can be sabotaged, such as *freeloading, cheating, defecting,* and *theft.*

Despite this, human societies have found ways to maintain large-scale coalitions, although sometimes at a severe cost, as we shall see. I imagine that most people reading this book run only the occasional risk of such unhappy events as being burgled or taken in by a used-car dealer, although I write too soon after the terrorist attacks on New York and Washington to be entirely sure of this. Nevertheless, Paul Bingham has argued that we have been at least moderately successful in preserving stable coalitions by becoming the first and only species to be able to kill at a distance.[43] We saw in the previous chapter that the early hominins probably evolved the ability to throw missiles accurately enough to maim or kill, or at least bring down an opponent so that more direct attack was possible. This was critical, according to Bingham, not because it enabled our forebears to kill other species— although they undoubtedly did this, and we still do—but rather because it enabled them to kill each other with relatively little risk to the killer! There can be little doubt that the evolution of deadly weaponry has been a trademark of our species—a stunning achievement, one might say, except that it goes well beyond mere stunning.

Consider the problem of a coalition faced with the presence of a freeloader. Let us suppose that the benefit of belonging to the coalition scores 5 on some scale, while the cost scores 3, giving a net benefit of 2. The freeloader does not pay the cost and so enjoys a net benefit of 5. In trying to exclude the freeloader, the members of the coalition incur some extra cost, perhaps in the form of death or injury. But they only have 2 points to "spend," as it were, since if the cost is more than 2 points they effectively lose any benefit of remaining in the coalition. The freeloader has 5 points to spend, and therefore can afford to take more risk. This means that the coalition is highly vul-

[42] *Dunbar 1992.*
[43] *Bingham 1999.*

nerable to the disrupting influence of freeloaders, which is why stable coalitions among nonkin are rarely found in nature, except in human societies. Moreover, given this instability, members of coalitions must always be tempted to defect, unless there are ways to achieve what Bingham calls "coalition enforcement."

Bingham argues that the cost of expelling a freeloader is much reduced if the members of the coalition have access to ways of killing or injuring from a distance. It is not entirely clear to me that this is the decisive factor, as Bingham maintains, since the cost is also reduced for each individual if it can be shared among members of the coalition. Ten lions can expel a single freeloading lion with relatively little cost to themselves, just as ten men with guns can exclude a single armed man with relatively little risk of themselves being shot. But in terms of coalition maintenance, are the men with the guns better off than the lions simply by virtue of being able to strike at a distance?

Perhaps so. One possibility is that coalitions lead to the invention of better weapons, and it is certainly true that much of hominin evolution can be described as an arms race. We have progressed, if that is the word, from simple rock throwing to spears, to bows and arrows, to guns, to conventional bombs, to nuclear weapons. And Bingham cites impressive statistics to the effect that humans are more ruthless in killing fellow coalition members than in killing traditional enemies. In the twentieth century, now mercifully behind us, at least 170 million people, and possibly as many as 360 million, were killed by their own governments, whereas the number who died in the great wars of the century was a "mere" 42 million. About one third of all Cambodian citizens were killed between 1975 and 1979 in the Khmer Rouge era.[44] These are perhaps extreme examples of the cost of maintaining large-scale coalitions, but in the words of Kurt Vonnegut, so it goes.

It may also be true that killing is easier the further removed one is from the act itself. The task of the judge who pronounces the death sentence may be easier than that of the hangman who attaches the noose. It is said that those who released bombs from aircraft in World War II felt little remorse compared with soldiers who shot to kill on the battlefield. Killing with a bayonet was probably even harder. Perhaps politicians have the easiest task. In *The Lathe of Heaven*, the American writer Ursula K. Le Guin wrote, "He had grown up in a

---

[44] *Scully 1979.*

country run by politicians who sent the pilots to man the bombers to kill the babies to make the world safer for children to grow up in."

Perhaps, though, we do not really need to view our species as quite so trigger-happy as Bingham implies. Coalitions can be enforced without a firing squad or its equivalent, and society has devised all sorts of rewards and punishments to ensure conformity. These include fines, imprisonment, detentions, banishment to Australia, or mere scoldings for those who stray, and bonuses, salary increments, medals, or mere words of encouragement for those who work hard for the good of their company or their country. Or, in rare instances, their university. One fairly universal and powerful system for enforcing coalition is religion, which promises eternal damnation for the wicked nonconformer and heavenly bliss for the compliant. The golden rule, "Do unto others as you would that they should do unto you," is a good recipe for cooperation.[45] And it's better than killing babies.

■ Coalition enforcement and the emergence of mind ■

The mechanisms of coalitional enforcement are, of course, heavily dependent on mental developments. The arms race itself is testimony to human inventiveness, perhaps even sharing some of the innovativeness of language itself. Bingham argues that the mental developments that underlie coalition enforcement actually stem from the ability to kill at a distance, but while this might provide a parsimonious account of human uniqueness, I suspect that the truth is more complex. The ways of enforcing coalition are surely too varied and subtle to be reducible to the firing squad, despite the approach all too often taken by military establishments. Or is it simply that I'm old enough to remember the 1960s slogan, "Make love, not war"?

In any event, it is clear that our mental processes are very much tuned to social situations, many of them having to do with cooperation and the enforcement of coalitions. In order to maintain coalitions and to recognize the dangers of noncompliance, our forebears needed to be able to understand what others saw or felt, heed warnings, and understand the options for response. A good example is our facility for "cheater detection."[46] This can be demonstrated with an adaptation of a simple

[45] This is a late-fifteenth-century proverb. A similar sentiment was expressed by Confucius around 500 B.C. When asked whether there was one word which can serve as a guiding principle throughout life, he replied "It is altruism. Do not do to others what you do not want them to do to you."

[46] Not to be confused with cheetah detection, which would also have been useful on the savanna.

test of reasoning, developed by the British psychologist Peter Wason.[47] Suppose you are shown four cards, bearing the symbols A, C, 22, and 17, and are told that there are also symbols on the other side of each card. You are then asked which two cards need to be turned over to check the truth of the following claim: "If a card has a vowel on one side, then it has an even number on the other side." If you're like most people, you'll choose the cards displaying the A (a vowel) and the 22 (an even number). It is indeed rational to turn over the A, but turning over the 22 is not really very revealing, since whatever is on the other side cannot disconfirm the statement. The better strategy is to turn over the 17, because the presence of an A on the other side would falsify the statement.

But now suppose it is explained that the symbol A stands for *ale* and C for *coke*, and it is explained that there are beverages on one side and people's ages on the other. If asked which two cards to turn over to check the truth of the statement: "If a person is drinking ale, he or she must be over twenty years old," most people easily understand that the critical cards are those bearing the labels A and 17.[48] This task is formally the same as that involving the letters and digits as abstract symbols, but now the policeman in you readily understands that you should examine what the seventeen-year-old is drinking if you want to stamp out underage drinking. Tooby and Cosmides argue from this that most of us are poor logicians but are nevertheless blessed with "cheater-detection" antennae that evolved during the Pleistocene era to ensure good social behavior. Whether cheaters were killed at a distance, though, is a moot point.

Cooperation also implies what has been called *theory of mind*. As explained in chapter 1, this refers to the ability to understand the minds of others or to see the world from the perspective of another person. This provides a basis for cooperation. If we share the feelings and knowledge of others, we are much more likely to help them; to help alleviate distress in another person is to alleviate one's own distress, even if it involves risk. Theory of mind can be a mixed blessing in other ways, as Orlando observed in Shakespeare's *As You Like It*: "Oh, how bitter a thing it is to look into happiness through another man's eyes." Although some have claimed that the chimpanzee is capable of theory of mind,[49]

[47] Wason 1966.
[48] Example adapted from Cox and Griggs 1982.
[49] As I noted in chapter 2, Savage-Rumbaugh (Savage-Rumbaugh, Shanker, and Taylor 1998) claims that the bonobo Kanzi displays theory of mind, and Byrne 1995 has inferred theory of mind in great apes from carefully analyzed instances of tactical deception.

this is probably true in only limited ways, and may often reflect little more than an innate ability to follow the gaze of others, as we saw in chapter 3. Simon Baron-Cohen has argued that true theory of mind was not present in the common ancestor of chimpanzees and humans 5 or 6 million years ago and probably evolved incrementally rather than as an all-or-none faculty.[50] I return to the relation between language and theory of mind in chapter 9.

Perhaps the most obvious way in which mental and emotional states can be shared, and coalitions enforced, is through language itself. Indeed, language and theory of mind are surely interdependent,[51] since we use language primarily to influence the minds of others, whether preaching hellfire and damnation or simply telling stories. As I pointed out in chapter 1, theory of mind has the same recursive structure as a sentence, as in *I know that she thinks I'm crazy.* Maybe you do, too. Moreover, language is of course fundamentally social and very often has to do with maintaining and enforcing cooperation. The preacher of hellfire and damnation is a good example, notwithstanding Sydney Smith's complaint of being "preached to death by wild curates."[52] It's true that people sometimes seem to talk to themselves, although sometimes, as in the case of prayer, they believe that someone out there is listening.

■ Was language gestural? ■

In this chapter, I have argued that the main characteristics of the human mind emerged over the past 2 million years. Except for the cultural component, human mental capacities had probably achieved their present level by the emergence of *Homo sapiens* in Africa around 150,000 years ago. The remaining question is whether language was primarily gestural or vocal during that period.

Although early *Homo sapiens* probably talked at least some of the time, I think there are good reasons to suppose that much of the development of language over the past 2 million years took place through manual gesture rather than vocalization. Firstly, as we have seen, our primate forebears were poorly equipped to generate intentional vocal signals, far better preadapted to making voluntary movements of the hands and

---

[50] Baron-Cohen 1999.
[51] See Premack and Premack 1994 for further discussion of this. There is some evidence, though, that syntactic language emerges in child development before theory of mind, although theory of mind may later facilitate further language developments (de Villiers and de Villiers 1999).
[52] From Lady Holland Memoir 1855.

arms. Secondly, voiced communication would have run the risk of detection, whereas manual gestures are silent. The !Kung San, who are present-day hunter-gatherers, sometimes use bird calls to communicate with each other while seeking prey, and then revert to silent signals as they advance on unsuspecting victims.[53] Thirdly, much of the communication would have involved locational signaling, indicating where predators or prey were lurking. Locational information is much more readily conveyed by pointing or by direction of gaze than by noises emanating from the throat.

The development of throwing and toolmaking skills would have added a further component of mime to the communication repertoire, leading to what Merlin Donald has called a *mimetic* stage in hominin evolution.[54] The actions involved in making or using tools could have come to represent the tools themselves, or perhaps the hands and arms were used to depict the actual shapes of things. Thus the hunting and killing of an animal or the making of a tool might be enacted as a mimed sequence, whether as a means of instruction or as a way of reporting a remembered event or planning a future one. Mimesis grew out of the primate ability to program sequences of action, and the freedom of the arms and upper body in our bipedal forebears would have lent greater variety and iconic quality to these actions.

According to Donald, though, mimesis does not constitute protolanguage but is merely a precursor to it. Its contribution to language was simply that it set the stage for the voluntary programming of vocal speech acts, which have in fact been called "articulatory gestures" by some researchers.[55] Language would then have evolved as a vocal accomplishment, while mimesis lives on in dance, mime, body language, ritual, some forms of music, and nonverbal communication. This is not my own view: I see no compelling reason *not* to count mimesis as protolanguage, since it involves combinable acts that can be arranged in different sequences. Mime and language may be less distinct than is commonly thought, a point that should become clear when I discuss signed language in the next chapter.

It is commonly held that nonverbal communication is somehow different from language, or that there is an independent form of language, called body language, that somehow tells what speech cannot. People characteristically gesture as they talk, and gesturing does indeed sometimes convey important information not carried by speech.

[53] *Lee 1979.*
[54] *Donald 1991.*
[55] *Donald 1999.*

For example, in describing the size of a fish she is alleged to have caught, a person may say, "Well, it was about this big," while holding her hands apart at about the length of the remembered victim. Or perhaps slightly further apart. Most people, when asked to describe a spiral, will resort to a manual demonstration. Saying something with raised eyebrows—a facial gesture—may substantially alter its meaning, as academics well know. The Italians seem particularly prone to gesturing while they speak.[56] I once tried to explain the gestural theory of language to a prominent linguist, who dismissed it with an eloquent wave of his hand.

"So you think gestures are nonverbal?" the psychologist David McNeill once asked, no doubt with raised eyebrows.[57] He has shown, on the contrary, that the gestures we use when we speak are in fact precisely synchronized with the speech, suggesting that speech and gesture together form a single, integrated system. He distinguishes between two kinds of gestures. Punctuating gestures, also sometimes called *beats* or *batons*, convey no meaning, but serve to give emphasis, as when a school teacher or football coach is laying down the law. Iconic gestures, also called *deictic* gestures, do convey meaning. McNeill and his colleague Susan Goldin-Meadow propose that speech carries the grammatical component, while iconic gestures help convey the actual content, especially if it includes spatial or emotional components difficult to put in words. If people are prevented from speaking while they explain something, their gestures then begin to take on the grammatical component as well.[58] Language is a son-et-lumière display, a blend of sound and sight. It is readily pushed in either direction, toward total gesture, as when we try to get a message across to those who speak a different tongue, or stripped of all gesture, when a message reaches us via telephone or radio. To be sure, speech dominates, but gesture is not far below the surface. And people still gesture when they *talk* on the telephone, or on radio.

There is one aspect of deictic gestures that suggests that they may have preceded vocalization. When people are required to gesture and vocalize at the same time—for example, by simultaneously naming a symbol that appears on a

---

[56] Critchley 1975 suggests that gesturing is especially prominent among the Neapolitans and Sicilians, who may have inherited it from the theatrical traditions of their Greek, Roman, and Carthaginian ancestors. He also mentions a theory that the use of gesture in Sicily derives from the dictatorship of Hiero, who prohibited public meetings: people signed to one another so that their conversations would not be monitored. My Italian friends, I should say, are not noticeably reticent in using speech.

[57] McNeill 1985.

[58] Goldin-Meadow, McNeill, and Singleton 1996.

screen and giving a learned handshape for that symbol—gesture and vocalization compete with each other. But it appears to be an unequal competition. Speaking is slowed down slightly by the requirement to gesture, but the gestures are not slowed down by the requirement to speak.[59] There is also evidence that illustrative gestures usually precede the part of an utterance to which they relate, and never follow it, and that gesturing can facilitate word finding.[60] These phenomena could be taken to mean that gesturing is more firmly established in the communication system, perhaps because it goes back further into our evolutionary past.

I noted in the chapter 3 that young children point before they speak. Once they learn the names of the things they point to, pointing itself seems to gradually disappear, but it is replaced by other forms of gesture. By adulthood about 90 percent of gestures consist of approximately equal numbers of beats and iconic gestures, and pointing accounts for only about 5 percent of gestures.[61] More importantly, perhaps, nearly all of these gestures are made during speech, indicating that gesturing is not an alternative to speech or a compensation for an inability to find words. Iconic gestures, in particular, are an integral part of the language process itself.[62]

The emergence of the genus *Homo* almost certainly heralded more sophisticated behavior, including toolmaking and migrations, both of which may have involved more extensive cooperation among individuals. I suspect that language was still largely gestural, at least for the earlier members of our genus, but elements of grammar may have begun to emerge, enabling more complex ideas to be expressed. It may well have been these added complexities that swelled our heads. But where did the grammatical component come from? The clue to that, I think, may be found in the nature of signed languages themselves. And that is the topic of the next chapter.

[59] *Feyereisen 1997.*
[60] *Butterworth and Hadar 1989.*
[61] *Mayberry and Shenker 1997.*
[62] *See Mayberry and Nicolaidis 2000 for a summary of work on gestures in children, including their own work on the importance of gestures in bilingual children while simultaneously learning French and English. These authors conclude that "iconic gestures may be a central feature in the development of languages, both from an individual and evolutionary standpoint" (195).*

"In this sign shalt thou conquer."
—Constantine the Great (A.D. C. 288–337)

6 ■ Signed Language ■

If our forebears communicated with gestures, we can perhaps gain some impression of what their language may have been like by examining present-day signed languages.[1] These have been around for a long time. Xenophon, writing in 431 B.C., mentions an encounter with signed language, and throughout history signing has been used by the deaf or in monastic communities. Girolamo de Cardano noted in 1576 that the deaf could express abstract ideas in signs, and in 1616 Giovanni Bonifaccio declared signing to be a universal language. The hypothesis that language itself originated in gestures, though, was first proposed in the mid-eighteenth century, by the philosopher Condillac. Then, as now, it met with opposition.[2]

Before 1750, 99.9 percent of those born deaf had no hope of becoming literate or educated. Things did not improve until the late eighteenth century, in France, when signed languages began to be recognized as legitimate languages. This was largely due to the influence of the abbé de l'Épée, who was determined to save the souls of deaf-mutes, hitherto deprived of the Word of God, by teaching them the Scriptures and the catechism. He was impressed with the animated way in which the deaf who roamed the

[1] I have adopted the convention of referring to languages that depend on manual signs as signed languages rather than sign languages. The term sign language will be used to designate individual signed languages, such as American Sign Language, British Sign Language, and so on.
[2] See Condillac 1947/1746 and, for a more detailed historical calendar, Hewes 1996. Condillac is no relation, by the way; our names just look a bit alike.

streets of Paris signed to one another. Echoing Condillac, he wrote: "The universal language that you scholars have sought for in vain and of which they have despaired, is here; it is right before your eyes, it is the mimicry of the impoverished deaf. Because you do not know it, you hold it in contempt, yet it alone will provide you with the key to all language."[3] Hear! Hear![4] The abbé set up a school for the deaf in 1755 and developed a combination of the deaf students' natural signs and signed grammar to teach them to read, and so receive an education. This establishment was the first school for the deaf to receive public support and by 1791 had become the National Institution for Deaf-Mutes, located in Paris.

It took some years for this enlightened attitude to spread elsewhere. Laurent Clerc, who had been a pupil at the institution, went to the United States in 1816 and immediately created an impression with his intelligence and remarkable erudition. With Thomas Gallaudet, he set up an American Asylum for the Deaf in Hartford in 1817. This established a strong tradition of teaching in signed language and led to the development of American Sign Language (ASL), which is a combination of the signs that Clerc introduced from France and the signs already in use among the local deaf population in the United States.[5] In 1864 the U.S. Congress passed legislation authorizing the Columbia Institution for the Deaf and Blind in Washington to become a national institution of higher learning for deaf-mutes. It was later named Gallaudet College after its first principal, Edward Gallaudet, son of Thomas. Now called Gallaudet University, it remains the only liberal arts university college in the world that caters exclusively to the deaf.

Yet then, as now, some had doubts as to the wisdom of using signs to educate the deaf and tried to overthrow the signed-language asylums (as they were then called) and replace them with "oralist" schools. This movement toward oralism gathered strength in the nineteenth century, encouraged by influential figures such as Alexander Graham Bell, who advocated the use of devices to amplify speech. Matters came to a head at an International Congress of Educators of the Deaf, held in Milan in 1880, when a vote in favor of oralism was passed, and signed language was declared officially prohibited. This attitude per-

[3] *Quoted in translation by Sacks 1991, 16–17. The historical material in the following four paragraphs is derived largely from Sacks's book, which is highly recommended.*
[4] *My apologies to the deaf for insensitivity, but I must applaud. I prefer, actually, the method of applause used by the deaf, which is to raise the hands above the head and flap them up and down. This gives the delightful impression of a flock of birds taking flight. Birds again.*
[5] *The influence of LeClerc also explains why ASL and French Sign Language have much in common.*

sisted until the late 1970s, with consequences that many feel to have been drastic. Surveys in the United States in 1972 and a few years later in Britain showed that deaf adolescents, on leaving secondary school, were on average able to read only at about the level of a nine-year-old.[6]

The tide began to turn again in the 1960s, largely through the efforts of the late William C. Stokoe,[7] a linguist and teacher at Gallaudet University. Even though signed language was not properly recognized even there, Stokoe saw that the students nevertheless used it virtually all the time in their informal interactions and that it had all the expressiveness of a true language. Due largely to Stokoe's influence, signed languages are now thoroughly redeemed as true natural languages, and American Sign Language (ASL) is now reestablished as the official language at Gallaudet. Students are taught all the usual subjects—mathematics, chemistry, philosophy, even poetry—without a word being spoken.

Most schools for the deaf in the United States have now switched from oral to visual communication systems. In some schools, though, teachers use an artificial signing system that follows the grammar of spoken English, and some educators favor teaching ASL as a first language and written English as a second. The learning of ASL in schools has also been hampered somewhat by the policy of "mainstreaming," introduced in the United States in the early 1970s, in which deaf children are placed in normal schools with hearing children, and so lack appropriate role models. In this environment, many deaf children come to regard ASL as somehow inferior.[8]

Arguments over the relative merits of oralism and signing are not merely about linguistic equivalence. Signed languages are inevitably confined to a tiny minority of the population, so those who can only communicate in sign are disadvantaged in that they cannot communicate easily with the great majority of people. Anyone who has visited a foreign country where they speak a different language will know the feeling. Moreover, although attempts have been made to develop written versions of signed language, these have not found general acceptance, and learning to read scripts based on spoken language is especially difficult to those whose only language is signed. There are therefore advantages in trying to teach deaf people normal speech, building on whatever limited hearing they may possess. Even so, one cannot but be impressed at

[6] Conrad 1979.
[7] See Stokoe 1960.
[8] See Neidle, Kegl, MacLaughlin, Bahan, and Lee 2000.

the ease and fluency with which deaf signers communicate with each other, relative to their often halting and painful efforts to speak or to understand the speech of others. In terms of sheer expressiveness, signed languages appear to be comparable to spoken ones. Some four thousand signs, for example, have been recorded in ASL, and this is probably a considerable underestimate of the total number.[9]

Signed languages are not confined to the deaf. Among the most intricate of signed languages are those invented by the indigenous Australians.[10] Apparently these originated in the North Central Desert of Australia and spread from there. They are, however, of comparatively recent origin and should therefore not be taken as direct evidence that vocal language derived from signed language. In fact, it is the other way about in this case, since the signed language is actually based on spoken language. Signing is used in part to overcome speech taboos, which are observed by women in the North Central Desert following the death of a close relative and are also imposed on male novices in initiation.

A system of signing was also developed in North America, probably before the arrival of Europeans. Known more recently as Plains Sign Talk (PST), it served mainly as a kind of lingua franca to allow tribes who spoke different languages to communicate with one another. It seems to have spread from the Gulf of Mexico and Southern Plains northward into the central and Northern Plains, and in the twentieth century into Saskatchewan and Alberta in Canada.[11] Detailed descriptions of PST began in the late nineteenth century, and one dictionary of signs published in 1880 lists over 3,000 signs.[12] It now appears to be restricted largely to tribal elders; English has replaced it as a lingua franca.[13]

Monasteries are another hotbed of signing. Many religious orders impose silence on their members, either completely or at particular times or in particular locations. The early monastic Fathers of the church held that silence was a prerequisite to a life without sin. According to Saint Benedict, for example, "in much speaking there is no escape from sin."[14] It is perhaps ironic that signed languages should be permitted, since we now know that signers can be as garrulous as speakers, although monastic signed languages tend to be deliberately pared down to restrict expression. In one list of signed languages drawn from the literature on monaster-

[9] *Crystal 1997.*
[10] *See Kendon 1988.*
[11] *Mithun 1999.*
[12] *Mallery 1880.*
[13] *Speaking from personal experience, Carol Patterson has assured me that it is still used quite widely.*
[14] *Quoted in Barakat 1987, 77.*

[15] *Something of the "pidgin" flavor of Cistercian Sign Language can be seen in the following rendering of the Lord's prayer (from Barakat 1987, 146):*

You me father stay God courtyard blessed b you name
You king courtyard come you w b arrange
This dirt courtyard same God courtyard.
Give you me this day you me day bread
Four give you me sin
Same you me four give sin arrange fault.
No arrange sin all same unload you me hard time
Four you courtyard power light four all time.

In this rendering, signs have been translated literally. Note that in places, the sense is conveyed by the sounds of the English equivalent, as in four give *for* forgive *and the signing of the letter B for the English* be.

For those who may not know, or who have forgotten, the Lord's Prayer in English goes something like this:

Our Father, who art in heaven, hallowed be thy name;
Thy kingdom come, thy will be done On earth as it is in heaven.
Give us this day our daily bread,
And forgive us our trespasses As we forgive those who trespass against us.
And lead us not into temptation, but deliver us from evil,
For thine is the kingdom and the power and the glory forever and ever.

[16] *See Stokoe 1987, who describes as "moonshine" the view that sign language originated in monasteries.*

[17] *Bradshaw 1997 has argued against the theory that language originated in manual gestures on a number of grounds, one of which is that "gesture really comes into its own in the context of pointing, alerting others, attracting attention, paralinguistic emphasis, and demonstrating how to do things" (102–3). He overlooks the power and expressiveness of sign languages. He also remarks that humans and apes share many gestures in common. To my mind, this is an argument for the gestural theory, not against it.*

ies, the total number of authorized signs varies from 55 to 472, although the monks often add new signs to compensate for the inadequacies of the original list. In case you should seek shelter in a Cistercian monastery, Robert Barakat lists 325 authorized signs of the Cistercian monks, along with about 200 derived signs made up by combinations of basic signs. *Angel*, for example, is designated by combining the signs for *wing* and *saint*, and *sugar* by the combination of *flour* and *sweet*. Many of the signs have an iconic or pictorial quality, but the grammar is primitive, based loosely on the spoken language of the monks. In France, word order is based on French, while in the United States and England it is based on English. But in both cases the grammar is simple, closer to that of pidgin than that of full-fledged spoken language.[15]

■ Are signed languages really true languages? ■

Some signed languages, like those invented by monastic communities, are probably closer to protolanguage than to true language, partly because they are not learned in early childhood as natural languages, and partly because of the artificially imposed restrictions on free expression. It was once thought that it was the monks who invented signed language and bequeathed it to the deaf, but if anything, the opposite is true.[16] There can now be little doubt that the signed languages that have emerged naturally in different deaf communities around the world have the spontaneity and expressiveness of true language.[17]

As with spoken language, signing conventions are to some degree transmitted across cultures. As we have seen, for example, ASL includes a number of signs introduced from France, and different signed languages within Europe also share common features. Yet signed languages have also sprung up anew in various cultures, and the deaf children of hearing parents, isolated in their own families, develop a form of signed language known as homesign. The apparently spontaneous appearance of signed languages suggests that human beings do indeed have an innate ability for language, and that cultural transmission is not a necessary ingredient.

In this respect signed languages may be unlike spoken ones, which can be arranged in family trees according to their relationships to one another. As we shall see in chapter 7, it has even been argued that all spoken languages can be traced back ultimately to a single, original language, sometimes called Proto-World, or "the mother tongue."[18] Whether or not this is true, the extensive relationships among spoken languages do at least raise the suspicion that spoken languages may be a cultural invention, handed down from generation to generation but subject to changes with the passage of time. This is perhaps why speech is sometimes regarded as what Richard Dawkins has called a "meme,"[19] rather than a biological instinct.[20] I think there may be some truth to this, as I shall explain in chapter 9.

Some researchers have found evidence that deaf children spontaneously add grammar to primitive forms of signing improvised by their hearing parents. In one study of four children raised in the United States and four in China, all of whom had developed sophisticated signed languages, the signing of the children across the two cultures had more in common than the signing of either set of children had with that of their parents.[21] This is effectively an example of *creolization*, in which children embellish a simple pidgin by adding grammar to it, as we saw in chapter 1. Another striking example of the creolization of a signed language comes from Nicaragua, which had no schools for the deaf until the Sandinista government took over in 1979 and reformed the education system. Although the children in the newly created schools for the deaf were drilled in oral techniques, they

[18] Shevoroshkin 1990.

[19] As explained in chapter 1, Dawkins 1976 coined the term meme, by analogy with gene, to refer to characteristics that are transmitted culturally rather than via genetic inheritance. Blackmore 1999 suggests that language might be a meme. My suggestion here is that speech has more memetic elements than signed languages do, although the distinction between memetic and genetic control is not an absolute one. Further, the cultural influence will depend on the circumstances. To the extent that children learning a signed language are taught some established signing convention, the memetic component is high, but children inventing homesign presumably rely more on an innate biological disposition.

[20] Pinker 1994 argues that language itself is an instinct rather than a meme, but makes no distinction between speech and signing in this respect.

[21] Goldin-Meadow and Mylander 1998.

nevertheless invented a system of signs, called Lenguaje de Signos Nicaragüense (LSN), which was basically a pidgin. But when young children began to join the school from about age four, the system changed so much that it is given a different name, Idioma de Signos Nicaragüense (ISN). In effect, ISN is a creole. It is more compact and sophisticated than LSN, and it has grammar.[22]

These examples might give the impression that there is a universal language of signs, but this is not the case. In signed as in spoken language, children may bring what Chomsky called *universal grammar* to bear on the language they learn, but actual signed languages differ greatly.[23] It has been estimated that there are as many as four or five thousand signed languages around the world.[24] Moreover, American Sign Language (ASL) and British Sign Language (BSL) are rather different from one another, even to the point of mutual incomprehension, despite the fact that the British and the Americans have (more or less) the same spoken language. This illustrates the fact that signed languages are largely independent of spoken languages, at least on the surface. The abbé de l'Épée, whom I quoted above, was clearly mistaken in declaring signed language to be *the* universal language. But an attempt has nevertheless been made to create a universal signed language, known as Universal Sign, perhaps dating from a banquet honoring the abbé de l'Épée in Paris in 1834. This has not been very successful, despite the fact that it is based largely on European and North American signed languages that are historically related and therefore somewhat similar to one another. Universal Sign was little influenced by more diverse forms, such as those of South America, Asia, or Africa.[25]

Of course, there may have been a universal signed language at some point in the evolution of our species. If my arguments concerning the gestural origins of language are correct, a group of hominins, perhaps a million years ago, may well have achieved a form of language with the full grammatical range and expressiveness of the modern signed languages of the deaf, and this could have have spread, with variations, to other communities, in much the same way that spoken languages diverged from a common source. My guess, though, is that gestural languages, then as now, lie closer in evolutionary terms to the biological adaptations that gave rise to grammar and the ability to represent objects and actions internally, and de-

[22] *See Pinker 1994, 36–37.*
[23] *Chomsky 1986.*
[24] *Hudson 1980.*
[25] *See Supalla and Webb 1995 for more detail on Universal Sign.*

pend less on cultural input. In a word, signed language may be more *natural.* I suspect, though, that early gestural language would have included vocal elements, although dominated by gesture. As already mentioned, it has been suggested that all present-day spoken languages may derive from a common spoken language, know as Proto-World, but this may represent, not the emergence of language itself but of the first language that was autonomously vocal, with gesture in a subsidiary role.

The way in which signed languages are acquired by children also highlights their naturalness. Indeed, it is often claimed that children exposed to a signed language from infancy learn it earlier and more easily than other children learn to speak.[26] This is based on the fact that children who learn to sign show evidence of signing a month or two earlier than children who learn a spoken language start using words. More careful analysis suggests, however, that the early signs used by the signers are not really equivalent to words, and even the budding speakers start by making more extensive use of gestures than of speech. By the time language has developed to the point where two-word combinations are produced, the speakers have switched to vocalization, while the signers combine pairs of signs. From that point on, the milestones in the development of language are essentially the same for speakers and signers.[27]

Deaf children raised by deaf parents who use signed language even go through a stage in which they "babble" in sign, making repetitive movements of the hands and fingers that parallel the *ga-ga-ga* type of vocalization one hears in normal infants exposed to speech.[28] Babbling is generally considered an important precursor to speech, but this remarkable observation suggests that it is more properly a precursor to *language*, whether spoken or signed. Yet normal children who are not exposed to signed language also seem to make babblelike gestures, which suggests that gesturing is as important as voicing in early childhood.[29] Regardless of whether the language a child subsequently learns is signed or vocal, it is the early gestures that provide the basis for reference, identifying the objects and actions to which names must be attached.

In summary, then, there are several indications that signed languages may lie closer to the origins of language than does speech. Even budding speakers begin by pointing, which

[26] See, for example, Meier and Newport 1990.
[27] Volterra and Iverson 1995.
[28] Petitto and Marentette 1991.
[29] Meier and Willerman 1995.

is necessary to indicate what words stand for—whether those words are ultimately signed or spoken. Moreover, signed languages have simply sprung up anew wherever there are deaf communities, whereas speech is arguably based at least partly on pedigree, handed down like standards of politeness, or the family silver, or political affiliation, from generation to generation. On the surface, at least, vocal language depends almost exclusively on convention, since spoken words of themselves convey virtually nothing as to their meaning. This is why the cultural component is more important for speaking than for signing. Although the signs of signed languages also become conventionalized over time, the ready ability of humans to manually mimic objects and actions means that signed languages are much more likely to emerge spontaneously.

Let's have a closer look, then, at what signed languages are actually like.

■ What are signs? ■

A *sign* in signed languages such as ASL is the basic unit that corresponds to the *word* in spoken languages. Signs are made by various movements of the hands on or near the signer's body, although expressions of the face also contribute. Some are produced with both hands and some with only one hand. Most signers prefer the right hand for one-handed signs, and for two-handed signs the right hand has the dominant or more critical role. Left-handers, though, may make signs round the other way; this is regarded as entirely appropriate and creates no confusion. In this respect, signing is not like writing, where the left-right direction makes a difference—if you start confusing *b*s and *d*s, as many kids do, you're in for a dab time. Although the hands make the basic signs, the head and face may be involved in signaling syntax, as we shall see below.

Are signs symbols? The influential Swiss linguist Ferdinand de Saussure argued that one of the critical features of language is the *arbitrary* nature of the symbols it employs. As I pointed out in chapter 3, the words we use typically bear no relation to what they represent. The word *duck* does not walk like a duck or quack like a duck. Indeed, it is *not* a duck. Nor even a *canard*. In this way, Saussure distinguished between the sequence of sounds that make up a spoken

American Sign Language	**Chinese Sign Language**	**Danish Sign Language**

■ **Figure 6.1.** ■

Different signs for *tree*.

word, which he called the *signifiant*, and what it represents, the *signifié*. This distinction is often blurred in signed language, where signs have a more iconic (or pictorial) relation to what they represent. This has sometimes been taken to imply that signed language is not a true language but is more like a mime show.

But some signed languages are more iconic than others. It has been claimed that the signs in Plains Sign Talk are about 98 percent iconic, which is perhaps understandable, given that it serves as a kind of lingua franca between tribes who do not have the same spoken language. The pictorial component is no doubt an aid to communication between those who may never have met before. Nevertheless, the seemingly iconic nature of signs is often an illusion, in that the pictorial aspect is only evident after one has discovered what the signs actually stand for. It's a bit like solving a difficult crossword clue; it's easy to understand the clue *after* the fact. Moreover, those signs that do have an iconic, or picturelike, relation to what they represent may vary quite considerably from one signed language to another. Figure 6.1 shows, for example, the sign for *tree* in three different signed languages. They are different from one another, yet each may be said to convey something of the actual shape of a tree. The signed languages of the deaf characteristically include signs that are not obviously iconic. In BSL, the sign for *sister* is a bent finger touching the nose, and perhaps only a Freudian would see iconic meaning in that.

Iconicity and arbitrariness are not opposites, but rather the ends of a continuum. As Charles Hockett put it: "To the extent that a symbol or system is not iconic, it is arbitrary."[30] It has long been noted, in fact, that signs tend to become less iconic and more arbitrary over time. Charles Darwin quotes a passage from a book published in 1870: "[The] contracting of natural gestures into much shorter gestures than the natural expression requires, is very common amongst the deaf and dumb. This contracted gesture is frequently so shortened as nearly to lose all resemblance of the natural one, but to the deaf and dumb who use it, it still has the force of the original expression."[31] For example, the sign for *home* was once a combination of the signs for *eat*, which is a bunched hand touching the mouth, and the sign for *sleep*, which is a flat hand on the cheek. Now it consists of two quick touches on the cheek, both with a bunched handshape, so the original iconic components are effectively lost.[32] Studies of deaf children inventing their own homesign also suggest that signs are initially coined for their resemblances to what they represent but are later adapted to a more arbitrary form.[33] Iconic resemblance may be necessary to get signs up and running, so to speak, but loses its importance once the signs are established.

As we saw in chapter 3, the switch from iconic to arbitrary signs is called *conventionalization* and applies fairly generally to communication systems. Once a sign has become conventionalized, the receiver can no longer rely on its resemblance to real-world objects or events to figure out what it means. The anthropologist Robbins Burling remarks that "Conventionalization represents, in part, the victory of the producer (signer, writer, speaker) over the receiver (reader, listener),"[34] but there must, of course, be agreement between producers and receivers if people are to understand each other. With conventionalization, communication moves into the cultural domain. It may appear that having to learn arbitrary symbols is merely a nuisance, but in fact it has powerful advantages.

For a start, arbitrary signs are typically shorter, as we saw above. This means that communication is more efficient, since the signs can be delivered more rapidly. To painstakingly create an iconic image of a battleship, or a cathedral, or even an actress and a bishop, would simply slow down the story and perhaps ruin the punch line. And of course many of the concepts we communicate

---

[30] Hockett 1960b.
[31] Quoted from W. R. Scott's The Deaf and the Dumb, 2d ed. 1870, 12, in Darwin 1965, 62.
[32] This example, and a historical survey, are from Frishberg 1975.
[33] See Morford, Singleton, and Goldin-Meadow 1995 for more extensive discussion and analysis.
[34] Burling 1999, 335.

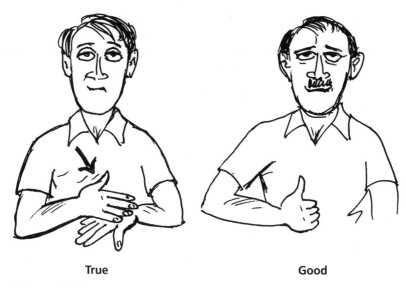

True   Good

■ **Figure 6.2.** ■

Signs for *good* and *true* in British Sign Language.

about are abstract in the first place, and the signs representing them cannot be decoded from the gestures comprising them. In British Sign Language (BSL), for example, the concept *good* is conveyed by the fingers closed in a fist with the thumb held upward, and *true* by the flat right hand placed, pinkie side down, on the left hand, flat palm upward, as shown in figure 6.2. Further, the sign for *good* in BSL means *male* in Japanese Sign Language, and the sign for *true* means *stop* in American Sign Language.[35]

But conventionalized signs have a more important advantage than brevity. Iconic signs can lead to confusion between objects or actions that look alike. Ducks look like drakes, and it would be virtually impossible to make hand signs that would distinguish their shapes. It would be similarly difficult to distinguish horses from donkeys, Fords from Chevrolets, a tennis shot from a badminton shot, or even cats from dogs—our flexible hands and arms are good at imitation, or online drawing, but not *that* good. Indeed, there are advantages to having signs for similar objects be as *dissimilar* as possible, so as to eradicate the possibilities of confusion; the spoken word *cat*, for instance, sounds very different from the word *dog*.[36]

[35] These examples are from Deuchar 1996.

[36] It must be said, though, that the ASL signs for these words cleverly preserve something of an iconic component, while still creating clear differentiation. The sign for cat *involves closing the thumb and forefinger to the right of the mouth and pulling away to the right twice, to indicate whiskers. The sign for* dog *involves patting the thigh twice, then snapping the thumb and forefinger twice, as though bringing a dog to heel.*

A well-designed language makes use of maximum contrasts, so as to optimize the fidelity and clarity of the message.

Arbitrariness is more or less forced on vocal language, since there is relatively little scope for picturing the world, or the events in it, using the sounds that the voice can make. Speech is intrinsically linear, providing little opportunity to represent the three dimensions of space.[37] There are of course some acoustical manifestations of the world that might be captured in the speech signal, but in practice we tend to steer away from onomatopoeia; the words for the animals we know do not sound like those animals, as already noted. Words that sound the same but have different meanings, like *bear/bare*, or *sole/soul*, or *creak/creek*, tend to crop up in widely different contexts, so that confusion is minimized. Vocalization, therefore, comes as a device well suited to the fabrication of appropriately contrasting signs. I shall return to this point in chapter 9.

■ Duality of patterning ■

Another characteristic of true language is that it has grammar, which imposes structure on what we say. In chapter 1 I emphasized the role of grammar in determining how words are put together to form sentences; this is called *syntax*. But there is another level of grammar that determines how words themselves are created from sounds; this is called *phonology*. Together these two levels make up what is known as *duality of patterning*. As an analogy, consider a city. There are principles, known to architects and builders, that govern the construction of all the individual houses, office buildings, and the like that go to make up a city. But we also need city planners, to determine how the houses and buildings should be laid out, to ensure ease of access, efficient traffic flow, and so on. So it is with language; we need to build words, and then we need to organize them into sentences. We might then go further and organize those into narratives.

Although duality of patterning is considered a hallmark of true language,[38] it has also been described as "one of the most difficult problems of language evolution."[39] How did we evolve a system that is structured in such different ways at two

[37] *This is perhaps not quite true. We can speak more slowly, with appropriate variations in pitch and loudness, to convey the impression of size—"He is a B-I-I-I-G man"—for example, or speak more quickly to convey an impression of speed or lightness. But such examples in no way match our ability to convey the three-dimensional shapes of things with our hands.*

[38] *Hockett 1960a.*

[39] *Armstrong, Stokoe, and Wilcox 1995, 156.*

entirely different levels, one involving sound and the other symbols? I shall suggest below that the mystery is largely solved if language evolved from gestures, and not from vocalizations.

But first, we need to ask whether signed languages themselves have duality of structure. To answer this, let's look at phonology and syntax and ask if the same principles hold for signed language as for spoken language. I need first to explain how phonology works in spoken language.

Phonology

Spoken words are made up of *phonemes*, which may be defined as the smallest units of speech that can make a difference to meaning. For example, the difference in meaning between *bat* and *cat* depends on whether the first phoneme is the /b/ sound or the hard /k/ sound, just as the difference between *cough* and *cot* depends on whether the final phoneme is the /f/ sound or the /t/ sound.[40] In written English, as in most present-day languages,[41] the individual phonemes are represented by letters of the alphabet, although the English language has such mixed origins that the correspondence is far from perfect. George Bernard Shaw once complained that the word *fish* might as well be spelled *ghoti*: *gh* as in *tough*, *o* as in *women*, and *ti* as in *nation*.

Actually, each individual phoneme represents a variety of different physical sounds, and not just a single sound. For a start, individuals have distinctive voices, so we all utter the phonemes of our language differently. But the same individual also utters phonemes differently, depending on the context in which they are embedded. For example, in terms of the physical pattern of sound, the /f/ in *fish* is not the same as the /f/ in *coffee*, just as the /b/ in your *bonnet* is not the same as the /b/ in your *bed*. This is a difficult point to grasp, as we tend to hear individual phonemes as the same when they are not. Indeed, this was not really understood until the invention of a device known as a sound spectrograph, which provides visual displays of the sound-frequency bands of speech plotted over time. It transpired that many phonemes are simply not discernible on the sound spectrograph, even though we can hear them quite clearly, and that the same phoneme might have quite different spectrograms. Techni-

[40] I have adopted the convention of putting slashes on either side of symbols representing phonemes.

[41] There are exceptions, of course. In Chinese and Japanese Kanji, the symbols stand for whole units of meaning, known technically as morphemes. In Japanese kana scripts, the symbols stand for syllables, not phonemes.

cally, the actual sounds that we utter are called *phones*, and phonemes are really highly abstract categories of phones.[42]

The reason that phonemes vary physically between different words has to do with what is called *coarticulation*. The reason that the /b/ in your *bonnet* is not the same as the /b/ in your *bed* is that your lips and mouth have already formed to make the following /o/ sound (for the following /e/ sound), and this alters the way the sound actually emerges. Watch yourself in a mirror while you say *bonnet* and *bed*, and you will see that your mouth makes very different movements; the fact that you don't really hear the difference in the /b/ sounds is owing to a remarkable cover-up carried out by the brain. Actually, you can't articulate a /b/ sound properly at all unless it is followed by another voiced phoneme, so a phoneme is utterly at the mercy of its context. There can be especially striking differences between consonants, depending on whether they precede or follow a vowel. Consider, for example, the words *rob* and *rod*. If these words are uttered in isolation, or at the end of a sentence, the final phoneme is often not articulated as such but merely creates a slight modification of the vowel. Sometimes, I suspect, we can only tell the difference by watching the speaker's mouth, which closes at the end of *rob* but remains open at the end of *rod*. In words like *bog* and *dog*, in contrast, the /b/ and /d/ sounds are articulated much more distinctly. Phonemes, especially the so-called plosives /b/, /d/, /g/, /p/, /t/, and /k/, therefore have a phantom quality, to the point that one might wonder whether they really exist. It is because individual phonemes vary so much that it has proven so difficult to design computers that can take dictation. The human brain solves the problem beautifully, but no one knows exactly how.

Different languages vary greatly in the number of phonemes they use and in the phonemes themselves. English is made up of 44 phonemes, which is a bit above the average for languages in the world as a whole. This number does not, of course, correspond to the number of letters in the alphabet, since some of the phonemes are represented by combinations of letters, such as /ch/, /ng/, /sh/, or /th/. The Maori language makes do with only 15 phonemes. There is no evidence that this detracts in any way from its power or expressiveness, and Maori speakers are noted for fine oratory. The Khoisan language of sub-Saharan

[42] *We are of course quite adept at dealing with categories of entities that often don't have a lot in common physically. Think, for example, of the variety of dogs. Or worse, the varieties of cruelty.*

Africa uses 141 different phonemes, including the characteristic click sounds unique to the region. Mastering new phonemes is one of the main obstacles to fluency in another language, especially if we try to learn it as adults. For example, Japanese makes no distinction between the sounds of /r/ and /l/, which is why Japanese people have great difficulty with words like *parallel* or Dylan Thomas's "Llareggub." This is not a matter of race, in the biological sense. No matter who we are, we have no difficulty acquiring a language, *any* language, provided we are exposed to it from an early age.

Phonemes can be divided into consonants and vowels, and most words can be broken down into syllables comprising consonant-vowel or consonant-vowel-consonant combinations. Of course, consonants are clustered in some words, such as *constraint*, but I can think of no English word that doesn't have at least one vowel—except perhaps *Pssst*! or *tsk*! Japanese can be broken down entirely into consonant-vowel syllables, and the elements of Japanese katakana script represent these syllables, not phonemes. It has been suggested in fact, that speech originated in consonant-vowel syllables, as in the babbling of an infant, such as *ga ga ga*, or *ba ba ba*,[43] with more complex variations coming later, but that's for the next chapter.

We can also identify phonemelike elements in signed languages, and linguists who deal with signed languages use the term *phonology* to describe these elements, even though they are silent! An important difference, though, is that the elements of speech occur in sequence, while in signed language they can be available simultaneously.[44] Individual signs are usually made up of three different kinds of components, not two as in the consonants and vowels of spoken language. In the system devised by William C. Stokoe and his colleagues, these are given the peculiar names of *tab* for location, *dez* for handshape, and *sig* for movement.[45] In BSL, for example, the sign for *know* consists of touching the forehead with the thumb, with the fingers closed. Here, the tab is the forehead, the dez the thumb extended from the closed fist, and the sig the movement to touch the forehead. The different possibilities available for each tab, dez, and sig are equivalent to phonemes and constitute the "phonology" of signed

[43] MacNeilage 1998.

[44] This distinction is not quite absolute. Phonemes overlap to some degree in speech; as we saw earlier, each phoneme is modified by its context, so the oo of boot makes its presence felt (or, rather, heard) even while you're uttering the b. And in signing, the tab (location) may not be fully identified until the sig (movement) is completed, as when one moves the hand to touch the forehead.

[45] Stokoe, Casterline, and Croneberg 1965. Using this terminology, these authors developed a system for representing signs in print, but it has not been widely adopted.

language. As in speech, different signed languages have different sets of elements. For example, in BSL the cheek and ear represent different tabs; the sign for *cheeky* is to hold the cheek between the thumb and index finger and shake it to and fro, while the sign for *lucky* is to do the same with the ear. In ASL, by contrast, there is no distinction between the ear and the cheek, just as in Japanese speech there is no distinction between /r/ and /l/. The sign for *deaf* in ASL used to be a touch on the ear, but is now a touch on the cheek—another example, incidentally, of the loss over time of the iconic component.[46]

The "phonemes" of signed language, like those of speech, are also coarticulated, and therefore vary with context. People's voice qualities differ, which gives their phonemes different acoustic properties, but people also differ in physical shape and size, so that the same gesture varies from person to person. Ways of gesturing may, indeed, be as distinctive as voices. I have brothers who are identical twins, and most people have difficulty telling them apart, but having grown up with them, I can easily distinguish not only their voices but also the ways they move, whether walking, playing tennis, or gesturing.

The actual movements (sigs) that are made in signed language are influenced by the previous movement, and by the following one, just as phonemes are influenced by neighboring phonemes. Hand-shapes also depend on preceding and following handshapes. Besides this, the actual view of a handshape will depend on where the hand is in relation to the viewer, and this will depend on the particular gesture the handshape is part of. Again, though, our clever brains are adept at seeing the constancies in a changing environment—a faculty not unique to our perception of signed language. We are able to see individual objects—shoes, ships, sealing wax—regardless of whether they are near or far, upright or askew, in sunlight or shadow, despite the virtually infinite number of ways they can be projected onto our retinas.

The study of signed language may actually give us a rather different perspective on the formation of words, whether spoken or signed. It has been suggested that spoken words might themselves be better understood as gestures, rather than as collections of phonemes. Some phonemes, at least, have little acoustic reality at all and may even be an artificial product of literacy.[47] It may be more appropriate

[46] The examples in this paragraph are from Deuchar 1996.
[47] Warren 2000 has recently argued that phonemes play no role in the perceptual processing of speech. One wonders if phonemes, as psychological entities, might be doomed for extinction, like phlogistin or the ether.

to think of speech, not in terms of combinations of those phantom entities called phonemes, but rather as combinations of sound "gestures" that we can make by the deployment of six independent "articulators" in the vocal tract. These are the lips, the blade of the tongue, the body of the tongue, the root of the tongue, the velum (or soft palate), and the larynx (or "voice box"). By combining these in various ways, we can produce words.[48] Similarly, the hands and body can be deployed in different combinations to produce signs. And it should be clear that, in the evolution of our species, the gestural elements required to produce the necessary variety of signs were in place long before the elements required to produce words.

Syntax

It is also clear that all of the elements of syntax that occur in speech have their counterparts in signed languages. The study of signed-language syntax is relatively recent and has been hampered somewhat by the dearth of linguists who are fluent signers. Nevertheless, sophisticated accounts are beginning to emerge.[49] The fact that general theories of syntax have proven adaptable to signed language provides some support for Chomsky's notion of universal grammar—a grammar that applies to *all* languages, whether spoken or signed.

One difference between speech and signed language is that, in signed language, syntactic marking is often carried out simultaneously with the rest of the message, just as elements of individual signs are conveyed simultaneously rather than sequentially. Consider how we turn a sentence from affirmative to negative, for example. In English, a negative sentence is marked by the insertion of *not* into the sentence, often with other changes. *The cow jumped over the moon* becomes *The cow did not jump over the moon*. Or *I kid you* becomes *I kid you not*. Believe me. In ASL, as mentioned earlier, negation is signaled by shaking one's head, along with a furrowing of the eyebrows, as one

[48] This point of view has been eloquently expressed in Liberman and Whalen 2000. The late Alvin Liberman had long stressed the articulatory component in speech perception. The idea that speech is made up of gestures was developed earlier by Browman and Goldstein 1991, who note that their theory was based on one previously developed to describe skilled motor actions in general and in its preliminary version was "exactly the model used for controlling arm movements, with the articulators of the vocal tract simply substituted for those of the arm" (314). They do not, however, suggest that speech may have originated in manual gestures. Their view challenges the popular notion (e.g., Liberman and Mattingly 1985) that speech involves a separate "module," operating according to its own rules and functioning independently of other modules. Nevertheless, it is consistent with Liberman's view that speech perception *is* special, since speech perception involves the recovery of the gestures that go to make up a heard speech sound. Perception of other patterns of sound would operate according to different principles.

[49] Chomsky 1995. The most up-to-date theory of sign language syntax is probably that presented in Neidle et al. 2000.

signs the otherwise affirmative sentence. By shaking one's head while signing *go*, the meaning shifts from *I am going* to *I am not going*. Other syntactic markers can also be signaled with facial gestures. The sign sequence *cow-jump-moon* becomes a question, *Did the cow jump over the moon?* if accompanied by a forward movement of the head and shoulders and a raising of the eyebrows. If accompanied by a raising of the eyebrows and upper lip, with the head tilted back, the same sequence, *cow-jump-moon,* would be understood as a relative clause within a sentence, as in *The cow that jumped over the moon broke a leg.*

We often use similar facial expressions when speaking. For example, raising one's eyebrows can sometimes change an assertion into a question, as in *You're going out with **him**?* although a change of intonation may be required as well. Speakers sometimes alter facial expression or the tilt of the head when inserting clauses into sentences. There is indeed a rather familiar look to some of the syntactic devices used in signing, but they have tended to be ignored in studies of vocal syntax. This need not mean that the gestures came first (although I like to think they did), since it is possible that the facial gestures used in ASL derive from those that accompany speech.

Of course, syntax is not simply a matter of facial expression. Some syntactic markers depend on where the hands move in space or on the order of signs. The use of space is especially interesting and relies in part on a powerful device used in ASL to specify people and objects that are to be the topics of a discourse. That is, one can "set up" a conversation by identifying individuals or objects with different regions in front of the body. For example, one might sign *Bill* while pointing to a specific region to the right of the midline, and *Hillary* while pointing to a region to the left. (Remember them?) Thereafter, simply pointing to Bill's patch identifies Bill, and pointing to Hillary's patch identifies Hillary. This is roughly equivalent to using pronouns in place of the actual names of people or objects but is much more flexible, since several individuals can be kept "alive" at the same time and maintained throughout a conversation. Once the first speaker has identified the referents in this way, another speaker can point to the same regions to refer to the same objects. The first person (*I, me*) is indicated by pointing to oneself. This use of space is sometimes used in speech, albeit in more tangible fashion, as when the coach picks the team for the current play. "I'll take you,

you, you, you, you, and you," she says, pointing to each of the favored warriors in turn.

In fact, associating people or objects with locations in space is not confined to signed languages but is quite a human specialty, as is illustrated by a well-known memory device known as the *method of loci*. This has a long history, beginning with an anecdote about the Greek poet Simonides, told by Cicero in his *De oratore*. It seems that Simonides went to a banquet given by a nobleman named Scopas, where he was to read a poem in honor of his host. In the poem, he also included a passage praising Castor and Pollux. As a result, Scopas undertook to pay only half of the promised fee, telling Simonides that he must obtain the balance from Castor and Pollux. A little later, Simonides was called out of the banquet hall to meet two men, but on going outside found no one there. But while he was out the roof of the hall fell in, killing everyone inside. By luring Simonides outside so that he would escape death, Castor and Pollux had handsomely rewarded him for including them in his poem! The corpses in the hall were so badly mangled by the tragedy that they could not be identified, but Simonides was able to remember who had been there by picturing in his mind where they had all been sitting. So began the method of loci, which was actually developed as a formal system by Greek, and later Roman, masters of rhetoric.[50]

For myself, I often find I can best remember who was at a dinner party, say, by picturing the table and identifying people by where they sat. I wonder, then, if this facility relates to a period in the evolution of gestural language when locations were associated with people or objects under discussion and retained *in imagination* for the duration of the conversation. The method of loci may be a legacy of our gestural past.

But of course signed language involves time as well as space. As in English, sentences in ASL normally proceed from subject to verb to object, which is to say that ASL is an SVO language.[51] One can therefore tell which sign is the subject and which the object by noting where each occurs in a signed sentence, so that a sign sequence *whale swallow Jonah* means something different from *Jonah swallow whale*.[52] The interplay between time

[50] *Yates 1966.*

[51] *Some languages, such as Japanese, place the subject first, then the object, then the verb, and so are called SOV languages. About 75 percent of the world's spoken languages are either SVO or SVO. A further 10–15 percent, including Welsh, are VSO (see Crystal 1997).*

[52] *There are situations in which word order can vary, which leads some authors to argue that the order of words in ASL can be freely altered without changing meaning. This is refuted by Neidle and his colleagues 2000, 59–61, who point out that variations are marked by other tags and may serve to topicalize a given noun. For example, a signer may sign "John Mary love" with a tag associated with the sign for John to indicate a meaning along the lines of "It is John that Mary loves."*

and space can be illustrated by the verb *give,* which is expressed in ASL by holding the hand with the thumb and fingers closed, moving it from the location of the subject to the location of the recipient (indirect object), and then signing the item that is given (direct object). In moving the hand from Bill's patch of space to Hillary's, and then making the sign of a rose, one would be asserting that Bill gives Hillary a rose. The way in which locations in space are signaled can also indicate various other features. For example, placing an open palm toward a location indicates possession (*belongs to Bill*). Pointing to a precise location in space indicates a definite referent (e.g., *Bill*), while a general area in space might represent an indefinite referent (e.g., *someone*). In signing *Bill gives someone a rose,* the hand would start precisely on Bill's location, and then move to a more general area of space, with the fingers spreading as the hand moves to this area. The spreading of the fingers may be considered an instance of the verb *agreeing* with the indirect object.

Past and future are represented in ASL by an imaginary time line, which locates the past behind the signer, the present close to the signer's body, and the future in front of the signer. The sign for *yesterday* involves closing the fingers and extending the thumb, with the thumb first touching the cheek and then moving back along the jaw line to the ear. The sign for *tomorrow* starts the same way, but the hand is moved forward, with the wrist pivoting down so that the thumb ends up facing forward. *Future* is signed by holding the open hand by the head with the thumb up and palm facing inward, and then moving the hand forward. The further the hand moves, the further into the future is the time period in question.

Many more subtleties of signed-language syntax could be cited, but the examples I have given should give something of the general flavor. It seems that virtually every aspect of syntax as identified in spoken language finds its counterpart in ASL. Moreover, the use of space in ASL, both to locate people and to indicate time, gives the syntax an analogue quality that is missing from the syntax of speech. For example, in English we can distinguish first, second, and third persons (e.g., *I, you, she*), but in ASL any number of "persons," at least in principle, can be dotted around and referred to simply by pointing. And time duration can be represented on a continuous scale, whereas in English we are restricted to such phrases as *in the future, in the distant future,* and *in the far distant future.*

I noted earlier that signs representing objects tend to lose their iconic aspect over time and become abstract and conventionalized. This does not appear to be true of the syntactic use of space and time, which remain strongly iconic.[53] This grounding of syntax in the four-dimensional world of space-time may remove at least some of the mystery surrounding the evolution of syntax—on the assumption, of course, that syntax emerged in the context of gesture, not of speech.

■ Signed language and evolution ■

If signed languages did indeed precede spoken languages, we have no reason to believe that they would have resembled present-day signed languages, at least in their particulars. As we have seen, all signed languages are different, and there is no evidence that they descend from a common "Proto-Sign," in the way that spoken language is believed by some to have descended from a common "Proto-World." It would therefore be misleading to pick any individual signed language, such as ASL, and imagine that this might have been the way in which *Homo ergaster* or *Homo erectus* communicated. Yet the spontaneity with which signed language emerges in deaf communities suggests that it does indeed spring from an innate disposition that was implanted at some point in hominin evolution. The spontaneous signed languages of the deaf are clearly beyond the sort of protolanguage that chimpanzees and bonobos appear to be capable of and beyond the repertoires of ape gesture observed in the wild.

One question, of course, is whether the language instinct, if that is what it is, is equally disposed to manifest itself vocally or manually, as Chomsky has recently maintained,[54] or whether it did indeed originate in manual gestures. For reasons I have already made clear in the preceding chapters, I think that the facts of primate evolution favor an origin in manual gestures. But is there anything about signed language itself that might favor this argument?

Well, as I anticipated earlier, signed language might hold the answer to the mystery of duality of patterning. David F. Armstrong, William C. Stokoe, and Sherman E. Wilcox, in their book *Gestures and the Nature of Language*, point out that in signed language, individual signs have the same

[53] I am indebted to an unnamed (but left-handed) publisher's reader for pointing this out to me.

[54] In his essay "Language as a Natural Object," Chomsky 2000, 121, said: "Though highly specialized, the language faculty is not tied to specific sensory modalities, contrary to what was assumed not long ago. Thus, the sign language of the deaf is structurally very much like spoken language, and the course of acquisition is very similar."

basic structure as a sentence. Indeed, the distinction is sometimes blurred. The authors ask the reader to imagine swinging the right hand across the body and grasping the upraised finger of the left hand. This gesture might be interpreted equally as a verb, *to grasp*, or as a sentence, *I grasp it*. Armstrong and his colleagues argue from this illustration that the structure of the sentence is derived from the sign itself; that the sign is the *seed* for syntax.[55]

Here is another example.[56] Imagine a scenario in which a third person sees you and a companion together, and then leaves for a moment. When he returns, he shows surprise at seeing you alone. You immediately interpret that expression of surprise and make a gesture that indicates, "She went out there." Your gesture actually tells more than that, because it shows which way she went, and perhaps the expression on your face also conveys your own attitude toward this event. The gesture is a simple sign, but it is also like a sentence, in which your hand stands for your now-departed companion, and the movement of the hand describes what she did and the direction she went in. Again, it is suggested, we have in a single gesture the basis of syntax.

Many other examples can be drawn from everyday life. One is a gesture much used by the late Pierre Trudeau, former prime minister of Canada: the shrug. This eloquent gesture—raising the shoulders, spreading the hands, and raising the eyebrows effectively says, "Who knows?" Charles Darwin observed that the English shrug far less frequently than the French or Italians. Perhaps it is simply considered un-British: in Shakespeare's *Merchant of Venice*, Shylock says

Signor Antonio, many a time and oft
In the Rialto have you rated me
About my monies and usances;
Still have I borne it with a patient shrug.[57]

Another example is a dismissive waving of the hands that says, in effect, "Forget it." No doubt the reader can think of other examples, where an expression of the face, a movement of the hands, a tilt of the head conveys a simple message. Again, gestures like these, which we see everyday, can be interpreted either as signs or as simple sen-

[55] Although Armstrong, Stokoe, and Wilcox 1995 take this as an argument for the gestural origins of language, Carstairs-McCarthy 1999 has proposed that the structure of sentences is derived from the structure of spoken syllables. This idea, of course, is more in line with the idea that language evolved from the start as speech, rather than gestures. I find this unconvincing, because syllables typically do not convey meaning by themselves, whereas individual gestures do.

[56] This example is from an unpublished manuscript by the late William C. Stokoe, which he was kind enough to send me before he died.

[57] Act 1, scene 3.

tences. Do we need to look further for evidence of the origins of the sentence itself?

The idea that language evolved from gestures, therefore, takes much of the mystery out of duality of patterning, since both words-as-signs and sentences are built from basic gestures. Beginning with the sorts of simple gestures described above, it does not take much to add handshapes to represent different objects and hand movements to represent different actions. Indeed, the chimpanzee signs tabulated in chapter 3 might also be understood as simple sentences, but there is no evidence that the chimpanzee has elaborated them in the way that ASL signers have. In signed languages, individual signs are already generative, in that they can be constructed from different combinations of the basic elements—tab, dez, and sig—and the next level of generativity, the combinations of the signs themselves, would follow from this. My guess is that the precursors of *H. sapiens* had in fact evolved a form of signed language similar in principle, if not in detail, to the signed languages that are today used by the deaf. But if this was so, then we now need to ask how, why, and when signs were largely, though not completely, replaced by spoken words.

7 ■ It's All Talk ■

Étienne Bonnot was born to a noble family in Grenoble, France, in 1714.[1] He may have been known to his mates as "Ettie," but probably not as "Fast Ettie"—he was a sickly child, with such poor eyesight that by the age of twelve he still could not read. In 1720 his family purchased the region known as Condillac, which Étienne later took as his title. As the family dullard, he was sent to study theology and was ordained as a priest, although he never practiced. Nevertheless, his ordination gave him the grand title of Abbé Étienne Bonnot de Condillac, although his mates now simply called him "Condillac." And so shall I.

Condillac was greatly impressed by the philosophy of John Locke and, according to some, became more Lockean than Locke himself. He became interested in how language might have originated, but since the prevailing opinion was that it came from God, and he did not want to offend the Church, he was obliged to invent a fable.[2] He imagined two children, a boy and a girl, who had not yet learned any language and were wandering about in the desert after the Flood. In order to communicate they used gestures. If the boy, for example, wanted something out of his reach, "he did not confine himself to

[1] Condillac's 1747 Essai sur l'origine des connaissances humaines was translated in 1756 by Thomas Nugent, as An essay on the origins of human knowledge and is reproduced in facsimile in Condillac 1971. The biographical details are from the introduction to that book by Robert G. Weyant.

[2] Today, it might be fear of Chomsky, rather than of God, that would lead one to construct such a fable.

cries or sounds only; he used some endeavours to obtain it, he moved his head, his arms, and every part of his body."[3] These signals were understood by his friend, who was only too willing to help. Eventually there grew "a language which in its infancy, probably consisted only in contortions and violent agitations, being thus proportioned to the slender capacity of this young couple."

Condillac goes on:

> And yet when once they had acquired the habit of connecting some ideas with arbitrary signs, the natural cries served them for a pattern, to frame a new language. They articulated new sounds, and by repeating them several times, and accompanying them with some gesture which pointed out such objects as they wanted to be taken notice of, they accustomed themselves to give names to things. The first progress of this language was nevertheless very slow. The organ of speech was so inflexible that it could not easily articulate any other than a few simple sounds. The obstacles which hindered them from pronouncing others prevented them from even suspecting that the voice was susceptible of any further variation, beyond the small number of words which they had already devised.[4]

But of course speech eventually won out: "In proportion as the language of articulate sounds became more copious, there was more need of seizing early opportunities of improving the organ of speech, and for preserving its first flexibility. Then it appeared as convenient as the mode of speaking by action: they were both indiscriminately used; till at length articulate sounds became so easy, that they absolutely prevailed."[5]

That just about says it all.[6] You are nevertheless invited to read on.

So yes, it's true what they say: people talk. The remarkable dominance of the spoken word over our lives is surely a defining feature of the human condition. Of course, we gesture too, but unless you are fluent in signed language, you will know that it is very difficult to get your message across using gestures alone. When traveling in a country

[3] *Condillac 1971, 172.*
[4] *Ibid., 174.*
[5] *Ibid., 175–76.*
[6] *Nevertheless Condillac could not entirely disregard his theological training, and finds numerous examples of the use of gesture in the Bible. For example, he quotes Ezekiel 37: "the prophets instructed the people in the will of God, and conversed with them in signs."*

where they don't speak your language, you may well resort to gesture, but communication is very limited, and you may curse the foreigners' inability to understand you—or your own lack of attention in foreign-language classes in high school. As we saw in chapter 5, people do gesture as they speak, and the gestures can certainly help one make a point, as it were, but they are not sufficient by themselves. Curiously, people often gesture when they speak on the phone or talk over radio, but listeners can understand quite adequately without seeing these gestures. One of the difficulties people have in accepting the gestural theory is that speech seems so natural and dominant that it is hard to believe we ever communicated in any other way.

We get a rather different impression, however, if we consider what the common ancestor of ourselves and the modern chimpanzee and bonobo must have been like. As we have seen, these modern great apes have poor control over vocalization but relatively sophisticated control over the arms and hands, and they have highly complex visual systems. They cannot be taught anything resembling vocal language, but they can be taught gestural or visual communication, at least to the level of what Derek Bickerton has called protolanguage and with at least some ability to combine symbols to form new meanings. Based on the primate evidence, the common ancestor of ourselves and our nearest relatives, the chimpanzee and bonobo, would seem to have been destined for a communication system based on gestures, rather than one based on vocalization.

Or if you don't believe in destiny, let me put it this way. Suppose you were to go back 5 or 6 million years and try to establish communication with that common ancestor. You would no doubt be faced with much the same problem as that faced by modern investigators trying to communicate with modern great apes, and you would no doubt resort to visual rather than vocal methods. Yet evolution somehow took a different course. What on earth can have happened to bring about such a dramatic change of direction, and when did it occur?

In this chapter I try to piece together what must have happened, first by looking backward through the prehistory of spoken language, and then by trying to trace the physical changes in hominin evolution that converted us from wild gesticulators to smooth talkers.

■ Listening to the past ■

Just when talking began in the evolution of our species is unclear. The earliest incontrovertible evidence for speech is little over a hundred years old: Edison's recordings! But we can be pretty sure people have been talking for longer than that. Even though I am an advocate of the gestural theory, I will not try to persuade you that Shakespeare's plays were performed in gesture, although Shakespeare certainly understood the power of nonverbal communication. In *Henry VIII*, Norfolk has this to say of Cardinal Wolsey:

> Some strange commotion
> Is in his brain; he bites his lip and starts;
> Stops on a sudden, looks upon the ground,
> Then, lays his finger on his temple: Straight,
> Springs out into fast gait; then, stops again,
> Strikes his breast hard; and anon, he casts
> His eye against the moon: In most strange postures
> We have seen him set himself.[7]

But it is words themselves that tell most about the history of spoken language. It is often easy to see how words in different languages are related and stem from the same source. For example, Latin gave rise to present-day Romance languages, including Romanian, Sardinian, Italian, French, Catalan, Spanish, and Portuguese. English also has many words of Latin origin; the very word *language* derives from the Latin word *lingua*, meaning "tongue." Languages mutate over time, and in different speech communities the languages gradually diverge and eventually become mutually unintelligible. By grouping languages together according to their similarity, one can begin to see where particular languages might have come from, and even begin to map out the course of human prehistory.[8]

A pioneer of this approach was Sir William Jones, a late-eighteenth-century British judge in India, who is reputed to have learned some twenty-eight languages.[9] He noted the strong affin-

[7] *Act 3, scene 2.*

[8] *In his book* The Origin of Language *1994, Merritt Ruhlen presents lists of words from different languages and invites the reader to group them by similarity, and so discover language families and superfamilies for themselves! Purists might argue that this approach is a little unfair, since the words are selected in such a way as to lead the reader to preordained conclusions, but the exercise is entertaining and informative. Another highly readable account, combining molecular and linguistic evidence, is the book* Genes, People, and Languages, *by Luigi Cavalli-Sforza 2000.*

[9] *Sir William is also credited with the following useful advice: "My opinion is, that power should always be distrusted, in whatever hands it is placed"—(letter to Lord Althorpe, 5 October 1782).*

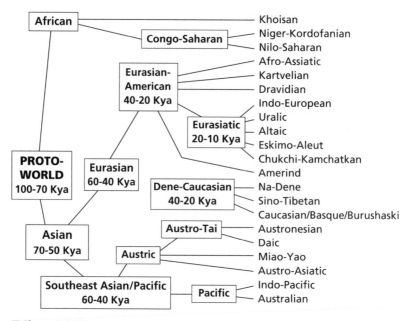

■ **Figure 7.1.** ■

A tree showing the possible evolution of language families. 1 Kya = 1,000 years ago.

ity between Sanskrit and Greek and Latin, which led him to identify a whole group of languages known as the Indo-European family. The common source of these languages probably arose in the region of the Danube somewhere between 5000 and 6000 B.C. Language diversity is greatest in Africa, and the American comparative linguist Joseph H. Greenberg identified no fewer than four African families, shown at the top of the list in the right-hand column of figure 7.1.[10] Khoisan is considered to be the oldest, perhaps dating back to the branch of *H. sapiens* that remained in Africa. The present-day Khoisan people now live in southern Africa, but there is evidence that they once lived further north, in eastern or northeastern Africa, and their ancestors may have been responsible for the spread of our species from Africa to Asia. The Khoisan family of languages all include the distinctive click sounds made famous by the singer Miriam Makeba in the 1960s. Another African family, Afro-Asiatic, has been attributed to a homesick population that migrated back to Africa, having previously migrated to Asia. The right-hand column of figure 7.1 shows a total of twenty-one language families accepted by the comparative linguist Merritt

Ruhlen (see below and note 8), but linguists disagree considerably as to the precise number and character of these families.

Some have argued that language families can in turn be grouped into superfamilies. For example, there is fairly general acceptance that Indo-European belongs to a superfamily known as Nostratic, although there is dispute as to which families belong to it. In Ruhlen's classification, it has been subdivided into Eurasian-American and Eurasiatic.[11] More controversially, though, it has been claimed that families and superfamilies can all be traced back to a single language, referred to as Proto-World,[12] or "the mother tongue."[13] Figure 7.1 shows the language tree proposed by Merritt Ruhlen,[14] based partly on linguistic but also on molecular evidence, and originating with Proto-World (see chapter 6) some 100,000 to 70,000 years ago.[15] Colin Renfrew suggested that theorists can be divided into the "lumpers," who argue for a single mother tongue, and the "splitters," who propose that languages cannot be reconstructed beyond about 5000 years ago. Ruhlen, along with his mentor Joseph H. Greenberg, is one of the lumpers, as are the so-called Russian school of Aron Dolgopolsky, Sergei Starotsin, and Vitaly Shevoroskin. With continuing molecular and archeological support for the view that all modern humans derive from a common African origin—the "African Eve" theory—the lumpers may be beginning to gain ascendancy.[16] Like it or lump it.

What do we know of Proto-World, or Eve's lexicon? There are a number of words that appear, with fairly minor modifications, in a wide range of the world's languages, which suggests that they might derive from our mother tongue. For example, variants of *mama* and *papa* occur in a great many languages; perhaps they each have a single common origin. One theory holds that the commonality derives simply from the fact that the sounds *m*, *p*, and *a* are among the first sounds

[11] Renfrew and Nettle 1999.
[12] Not to be confused with protolanguage.
[13] Two writers who have argued in an accessible way for a common mother tongue are Merritt Ruhlen 1994 and Vitaly Shevoroshkin 1990.
[14] This tree is based on that in Cavalli-Svorza 2000, which is in turn based on that developed by Ruhlen 1994. Ruhlen does not recognize Nostratic, which others believe to comprise the Eurasiatic group plus Afro-Asiatic, Dravidian, and Kartvelian. As Ruhlen admits, his own solution is also likely to undergo change with further investigation, but he is confident enough to assert that "large portions of it will prove valid even with the accumulation of additional evidence" (193).
[15] The date of 100,000 to 70,000 years is actually a sort of compromise between the dates suggested by mtDNA evidence, which is usually taken to indicate that H. sapiens originated some 150,000 years ago, and that suggested by linguistic mutation, which yields a much more recent date of perhaps 35,000 years ago. As we shall see, however, recent research on variation in the Y chromosome indicates that our most recent common ancestor may date from between 35,000 and 89,000 years ago, which is somewhat more compatible with the linguistic estimate.
[16] Renfrew 1992. He writes that ". . . there are already sufficient arguments fro the fields of archaeology, linguistics, and genetics to suggest that the new synthesis fulfilling the premonitory vision of Darwin and Huxley . . . may be more than a mirage" (473).

uttered by young children and become attached to the first objects that they learn to identify, namely, their doting parents. This theory does not easily explain why variants of the word *kaka*, associated with the meaning "older brother," also appear in diverse languages, since the sound *k* does not appear early in development.[17] Other so-called root words are *aqwa* ("water"), *tik* ("finger," or "one"), and *pal* ("two"), and minor variants of these words crop up in languages all over the world. Another possible root word is something like *nati*, or *natu*, meaning "nose," and persisting as *n!ati*, *!nutu*, and *!nasa* in various Khoisan languages (*!* is the click sound); *nici*, *nasa*, and *nus* in various Amerind languages; and of course *nose* and *nasal* in English. They must have been a nosey lot, since there appears to have been another word for nose, which may have sounded something like *changa* and persists, albeit transmuted, in the English words *snout* and *snot*.[18] It is possible that the click sounds of the Khoisan family might also be remnants from the mother tongue and may even derive from nonvocal click sounds made with the tongue, used to punctuate gestures prior to the emergence of speech.

The Proto-World hypothesis is still highly controversial among comparative linguists. But if all spoken languages are indeed derived from a single mother tongue, then it may not be unreasonable to suppose that spoken language is at least in part a cultural invention, handed down from generation to generation from the original population that invented it, as I pointed out in the previous chapter. My contention is that the capacity for autonomous speech arose in our own species, *Homo sapiens*, not in earlier hominins, although it may have taken some time for this capacity to be fully realized. Even today, of course, we have not fully escaped our gestural past, but we have developed the speech capacity to a point where we can converse over the telephone, or on radio, or with people deprived of sight and yet make ourselves fully comprehensible.

In order to see how this might have happened, let us examine more closely the emergence of our species.

■ Where did we come from? ■

The most recent evidence, based on mitochondrial DNA, not only continues to support the theory that all modern humans are de-

---

[17] *Ruhlen 1994.*
[18] *See Bengston 1992 for more detail and more examples.*

scended from a group of *Homo sapiens* who once lived in Africa, but suggests that our species was at first confined to a relatively small population, some 10,000 individuals, that lived about 170,000 years ago.[19] All of the evidence indicates that early *Homo sapiens* were essentially modern in form, presumably capable of understanding particle physics or the plays of Shakespeare, if only they had had the opportunity and experience. It can be reasonably assumed that they also had the capacity for autonomous speech, leading perhaps to the emergence of Proto-World.

It is unlikely, though, that Proto-World emerged overnight. Vitaly Shevoroshkin has argued that it was initially rather primitive: meaning was conveyed by consonants only, the only vowel being a short, gruntlike *a* sound. Vowels emerged later to help distinguish meanings; he suggests, for example, that the word *changa* initially referred to both the nose and to odor, but later on the word *chunga* came to refer to odor, while *changa* was retained for the nose. These events may well have taken place in the transition from a partly gestural system to an autonomous vocal system, perhaps in the period between 170,000 and about 100,000 years ago, or even more recently, before our garrulous forebears migrated from Africa.

There were almost certainly several migrations of *H. sapiens* from Africa, perhaps even a more or less continuous exodus. As we have seen, there is evidence of a migration along the coast of the Red Sea by 125,000 years ago, and then along a largely coastal route across Saudi Arabia, Iraq, and Iran to Pakistan, and from there along the Indian coastlines to Southeast Asia around 67,000 years ago. In perhaps several waves, our species had reached New Guinea by at least 60,000 years ago, and Australia by at least 45,000 years ago.[20] One group diverted to Europe, arriving there around 40,000 years ago,[21] and eventually replacing the Neanderthals. An Asian group crossed the Bering Strait into Alaska, reaching what is now the West Coast of the United States by around 20,000 years ago and South America around 13,000 years ago. Another Asian contingent ventured into the Pacific, reaching New Zealand by around A.D. 1300.[22]

But some evidence indicates that the descendents of the earlier migrants not have survived the

19 See Ingman, Kaessmann, Pääbo, and Gyllensten 2000 and Takahata, Lee, and Satta 2001. The latter are bold enough to assert that "the conclusion of a single foundation population of modern H. sapiens is inevitable".
20 See Cann 2001 for a summary.
21 Semino, Passarino, Oefner, Lin, Arbuzova, Beckman, De Benedictus, Francalacci, Kouvatsi, Limborska, Marcikiae, Mika, Mika, Primorac, Santa-Benerecetti, Cavalli-Sforza, and Underhill 2000.
22 Hedges 2000.

arrival of later ones, so the critical exodus that eventually gave rise to the present-day non-African population was relatively recent. An analysis of mtDNA gathered from fifty-three modern individuals from diverse regions of the world suggests that the nearest common ancestor of both African and non-African people lived as recently as 52,000 years ago, although the African lineages themselves go deeper, to some 170,000 years ago.[23] This suggests that there was an exodus from Africa around 50,000 years ago of a population that diverged and eventually replaced the indigenous hominin populations they encountered. These indigenous populations, destined for extinction, presumably included earlier hominins who had migrated—the Neanderthals in Europe and *Homo erectus* in Southeast Asia—and other *Homo sapiens*, descendents of earlier waves to leave Africa.

These dates have been largely corroborated by studies of variation in the Y chromosome. Just as mtDNA is passed on down the female line, so the Y chromosome is restricted to the male line and is not subject to recombination. Changes in the Y chromosome are therefore attributable to mutation only, and by considering the degree of variation and the rate of mutation one can infer the likely date of the most recent common ancestor of present-day men. One study of Y-chromosome variation among present-day men resulted in an estimate of 181,000 years ago for ancestral Adam,[24] which is in line with estimates for Eve, and places them happily together in the Garden of Eden, somewhere in Africa. But when the analysis is restricted to non-African men, the common ancestor is estimated to have lived somewhere between 35,000 to 89,000 years ago.[25] This is at least roughly consistent with an exodus from Africa around 50,000 years ago that gave rise to today's non-Africans.

It should not be altogether surprising that successive waves of migrants should replace earlier ones. Even in recent history, colonizers have had

[23] Ingman et al. 2000. These are only estimates and the common ancestor of Africans and non-Africans is actually dated at 52,000 years ago plus or minus 27,500 years; the most recent common ancestor of all modern humans, including Africans and non-Africans, at 171,500 years ago, plus or minus 50,000 years.

[24] Hammer 1995. The 95-percent confidence interval, though, was from 51,000 to 411,000 years ago, so perhaps Adam was indeed more recent than Eve.

[25] Underhill et al. 2000. This analysis also suggests that some contemporary East Africans and Khoisan are the descendants of Adam who remained in Africa. Others, however, may have ventured forth to conquer the world.

A more recent study by many of the same authors examined specific markers on the Y chromosome in 12,127 individuals drawn from 163 populations, mostly Asian. The common ancestor was almost certainly African and dated from between 35,000 and 89,000 years ago,—results that tell heavily against the regional-continuity theory that modern Asians evolved from much earlier migrants (Homo erectus) from Africa (Ke, Su, Song, Lu, Chen, Li, Qi, Marzuki, Deka, Underhill, Xiao, Shriver, Lell, Wallace, Wells, Seielstad, Oefner, Zhu, Jin, Huang, Chakraborty, Chen, and Jin 2001.

a devastating impact on indigenous peoples who had migrated earlier. The most extreme example in recent times is the indigenous population of the island of Tasmania off the coast of southeastern Australia. Tasmania was once connected to the mainland, but the Tasmanians had been effectively cut off from the aboriginal people of Australia for some 10,000 years before Europeans encountered them in the seventeenth century. Like the mainland Australians they were hunter-gatherers, but during their period of isolation they lost many of the technologies they had originally brought with them.[26] Consequently, they were unable to survive the subsequent European colonization. Yet there is no evidence to suggest that they were *biologically* inferior to other members of the species *Homo sapiens*.

Another possible example of replacement, to put it euphemistically, likewise comes from Australia. An ancient fossil, dubbed "Mungo Man" because he was found in the Lake Mungo region of Western New South Wales, has been dated at around 62,000 years ago, and the skeletal anatomy shows him to belong to our own species, *Homo sapiens*, rather than to *Homo erectus* or the Neanderthals. If all present-day non-Africans descend from migrations out of Africa beginning only 52,000 years ago, then this fellow must have belonged to an earlier wave that became extinct, presumably following the arrival of new migrants. Indeed, mitochondrial DNA extracted from this fossil is unlike that from any living human so far examined, including present-day indigenous Australians, and suggests a lineage older even than that of contemporary *Africans*,[27] which is perhaps something of a puzzle. The claims about Mungo Man are nevertheless controversial, and likely to remain so until further evidence on the early peopling of Australia turns up.[28]

[26] See J. Diamond 1997.

[27] See Adcock, Dennis, Easteal, Huttley, Jermiin, Peacock, and Thorne 2001. The mtDNA analyzed by these authors is by far the oldest mtDNA so far sequenced. With the advance of technology for sequencing ancient DNA, we can perhaps expect new revelations about our ancestry over the past 100,000 years. One may question whether evidence from such ancient DNA is reliable, but certain aspects of the analysis lend it credibility. One such aspect concerns a particular sequence in the mtDNA of Mungo Man that is found on chromosome 11 in living humans. This sequence jumped ship, as it were, from the mitochondria to the nuclear genome and is therefore referred to as an "insert" sequence. The fact that the sequences match seems good evidence that the sequencing of Mungo Man's mtDNA is genuine. The unique presence of the sequence in Mungo Man's mtDNA, rather than on chromosome II, strongly suggests that the common ancestor of Mungo Man and those carrying the insert lived before the common ancestor of all living humans.

It should also be noted that Adcock and his colleagues regard Mungo Man as evidence for the regional-continuity theory championed by Wolpoff (e.g., Wolpoff et al. 2001), whereas I have here suggested that his forebears may have been earlier migrants from Africa. This theory might be described as the multiple-wave theory.

[28] I happened to be in Australia when the conclusions about Mungo Man were announced, and greeted in the local press as a triumph for Australian science. But experts are already urging caution; Rebecca Cann, one of the pioneers in the use of molecular data to infer evolutionary trees, remarks that "Statistical confidence in their phylogenetic analysis is . . . questionable" (2001, 1747).

A comparison of evidence from mtDNA and Y-chromosome markers also suggests that women have dispersed their genes more widely than men.[29] This has been taken to mean that men tend to have children near where they were born, while it is the women who shift in order to be with their partners, at least in traditional societies. It can also be taken to mean "more rapid displacement of previous Y chromosomes," as one recent group of authors put it, relative to the rate at which mtDNA displaces earlier mtDNA.[30] This may tell something about sex differences in the way our ancestors carried on. The migrating—or perhaps I should say marauding—bands may have been made up largely of men, who plundered the earlier migrants, killing the men and abducting the women. But can we really believe that our grandfathers could have behaved so badly?

Steven Pinker seems to think so. He has gone so far as to suggest that, in foraging societies at least, the reason that men go to war is to obtain women: "The most common spoils of tribal warfare are women. Raiders kill the men, abduct the nubile women, gang-rape them, and allocate them as wives."[31] Not a pretty image, but one that appears even to have the blessing of the Lord:

> And they warred against the Midianites, as the Lord
> commanded Moses; and they slew all the males. . . . And the
> children of Israel took all the women of Midian captives,
> and their little ones, and took the spoil of all their cattle, and
> all their flocks, and all their goods. . . . And Moses said unto
> them, Have ye saved all the women alive? . . . Now
> therefore kill every male among the little ones, and kill
> every woman that hath known man by lying with him. But
> all the women children, that have not known a man by lying
> with him, keep alive for yourselves.[32]

This harsh view of our forebears is of course controversial and may be seen by some as providing a justification for the continuing ill-treatment of women—although it should perhaps be remembered that men were (and are) not treated so well by their enemies either. The women were raped, but the men were killed.[33] Further, as Pinker points out, it is better that we understand our nature, so that we can be equipped to protect ourselves from its excesses.

---

[29] *Seielstad, Minch, and Cavalli-Sforza 1998.*

[30] *Semino et al. 2000, 1159.*

[31] *Pinker 1997, 510.*

[32] *Numbers 31.*

[33] *Even so, it was the men who committed these evil deeds.*

Some of the evidence Pinker draws on is based on Napoleon Chagnon's controversial work among the Yanomamö, a group of Native Americans who live around the Orinoco River at the border between Brazil and Mexico and have, until very recently, remained almost entirely isolated from other human contact. The bloodthirsty nature of Yanomamö life, with stories of tribal warfare involving killing of men and the rape and abduction of women, is told in Chagnon's celebrated 1968 book *Yanomamö: The Fierce People*. Chagnon also claimed that, among the Yanomamö, men who kill are rewarded higher reproductive success; this group of hominins, at least, seems to strongly select for violence and killing.[34]

Be all this as it may, there are other indications that the decisive migrations for the future of our species out of Africa began only about 50,000 years ago. As we saw in chapter 5, the *Homo sapiens* who left Africa around 125,000 years ago, proceeding along the coast of the Red Sea, had tools that were Acheulian,[35] totally lacking in the sophistication and variety that was later to emerge. It is possible, then, that the successive waves of migration out of Africa were associated with increasingly advanced technology, causing each wave to succumb to the next.

The advance of technology may well have been largely due to the gradual emergence of autonomous vocal language. This may have been developed in Africa, perhaps between the appearance of anatomically modern *Homo sapiens* 170,000 years ago and the migrations that began about 52,000 years ago. That is, early *Homo sapiens* may have depended on manual gestures or, more likely, on a combination of gestures and vocalization in order to communicate. Indeed, judging from the rate at which spoken languages mutate, estimates as to when Proto-World would have emerged seem much more consistent with the more recent date of 50,000 years ago than with any earlier date. We must nevertheless suppose that earlier humans were *capable* of autonomous speech, since they seem to have been fully human anatomically. In short, autonomous speech may have been essentially a cultural invention, perhaps perfected by our African ancestors in the period prior to the decisive migrations of some 50,000 years ago. This is a matter I shall return to in chapter 9.

[34] Chagnon 1988. Chagnon's work has recently been strongly attacked in Darkness in El Dorado, by Patrick Tierney, who claims that the Yanomamö have been exploited and misrepresented by Chagnon and various associates.

[35] Walter, Buffler, Bruggemann, Guillaume, Berhe, Negassi, Libsekal, Cheng, Edwards, von Cosel, Neraudeau, and Gagnon 2000.

■ The anatomy of speech ■

Although early *Homo sapiens* was almost certainly capable of autonomous speech, it is likely that this capability emerged late in the evolution of the genus *Homo*, and it may even have been unique to our own species. One reason to suppose that it emerged late is that dramatic anatomical changes were necessary to convert our forebears from primates capable only of involuntary grunts and cries to fully articulate humans. We saw in chapter 2 that the vocalizations of other primates have little in common with speech. They are largely fixed and involuntary, with nothing like the flexibility and generativity of human discourse. Somewhere along the line from great ape to human we acquired the facility not only to utter an extraordinary variety of sounds but also to combine them in ever novel and meaningful ways, in order to produce what we are pleased to call speech. (You have never heard *that* sentence before, nor this one.) This breakthrough required extensive changes in the vocal apparatus, the way we breathe, and, of course, the brain.

Let's begin by examining the vocal tract.

How the Vocal Tract Changed

The structure that produces speech sounds is the larynx. It appeared first with the evolution of the lung and served to exclude everything but air from the pulmonary tract. It was originally a band of muscle around the glottis that closed the tract, much as one might tie a string around the opening of a balloon. Later, it evolved further, to prevent air from either entering or leaving the lungs, as when one holds one's breath. Later still, it evolved to create sound. Vertebrates from frogs to humans vocalize by passing air through the vocal folds, located in the larynx, which oscillate to produce sound. These successive functions of the larynx are classic examples of *exaptation*, whereby structures that evolved for one function are later borrowed for another.[36]

The basic frequency of the sound produced by the vocal folds can range from about 100 Hz (cycles per second) in adult men to about 500 Hz in small children. Although the oscillations of the vocal folds produce the basic timbre of a person's voice, the different vowel sounds, in particular, essentially depend on the way the sound is filtered by the vocal tract. In this filtering, other frequencies are created

---

[36] *Gould and Vrba 1982.*

as harmonics. The amplitude of some frequencies is much greater than that of others, and the peak frequencies are known as *formants*. For example, a man with a fundamental frequency of 120 Hz saying "ah" results in formants with peak frequencies at 360 Hz, 2,280 Hz, and 3,000 Hz. By altering the shape of the vocal tract, we can alter the formants to produce the full range of vowel sounds, from the highly constricted *ee* sound in *eek*! to the sheeplike *ah* sound in *baaa*! or the cowlike *oo* in *moo*![37] These three vowels, more formally represented as /i/, /a/, and /u/, are known as the "point" vowels, defining extremes in vowel production. They are the most common vowels found in human languages.[38] They are also somewhat analogous to primary colors, in that all other vowels lie within the "space" defined by them. Ask "Why?" very slowly and deliberately, exaggerating the vowel, and you will cover nearly all possibilities—or at least you will if you happen to be Australian or Cockney.

Figure 7.2 shows the vocal tracts of the chimpanzee and the human. The differences are obvious. In humans, the larynx lies much deeper in the throat than it does in the other great apes. This provides much more opportunity for modulation of the sound, since the vocal tract is essentially a right-angled tube that can be constricted at the angle point, that is, at the back of the mouth. This modulation depends critically on the so-called articulators, discussed more fully in the previous chapter: the lips, the tongue, the velum (or soft palate), and the larynx (or "voice box"). The tongue is the most important, which is why languages are often referred to as "tongues." In making the point vowels /i/, /a/, and /u/, it is really the tongue that does most of the work. For /i/, the middle-front portion is raised so that the sides touch the top teeth; for /a/, the whole tongue is dropped; for /u/, the rear of the tongue is raised and the lips are pushed forward. (Try making these sounds yourself to get a more accurate feel.) The tongue is critical to hard consonants like /k/ or /g/, which depend on closure at the back of the throat, where the angle between the two tubes is located. It is also involved in sounds like /s/, /t/, and /n/. The lips, too, are important and play the major role in sounds like /b/, /p/, and /w/. And teeth help; if you take them out, you will probably have difficulty with sounds like /f/, /v/, and /th/, and if you're not careful, you'll spit all over the place. Chimpanzees are simply anatomically incapable of making most of

[37] I have to cope with the possibility of different accents here, so I hope you will forgive my crude examples. But I'm sure you get the point.
[38] Jacobson 1940.

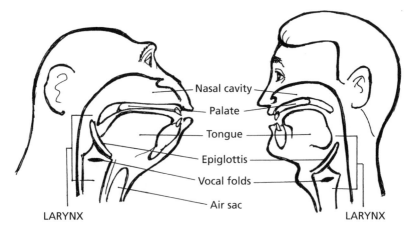

■ **Figure 7.2.** ■

Vocal tracts of chimpanzee and human, scaled to look the same size. Humans have a lower larynx and longer oral cavity, but the air sac is absent.

these vocal movements, which is one reason why they are virtually unable to produce anything resembling human speech.

The descent of the larynx into the throat may not have been the direct result of selection for articulate speech. Rather, it may have been a consequence of bipedalism. The spinal column enters the skull through an opening known as the foramen magnum, and in quadrupeds this opening is toward the rear of the skull. In bipedal humans, the foramen magnum is shifted forward and the skull tilted back so it can balance on top of the spinal column, resulting in a smaller jaw, a lengthening of the vocal tract, and a lowering of the larynx.[39] These changes would have occurred gradually in evolution as the bipedal stance was refined, perhaps reaching completion in *Homo ergaster* or *Homo erectus*, nearly 2 million years ago. If this account is correct, then the lowering of the larynx is an example of what has been called a *spandrel*[40]— a biomechanical consequence of a structural alteration. The alteration itself had nothing directly to do with speech, but accidentally happened to facilitate it.

The various distortions of the vocal tract that allow us to produce the different sounds of speech can be regarded as *gestures*, as I pointed

[39] *Buettner-Janusch 1966, 337. See also Armstrong 1999, 39.*
[40] *A spandrel is the architectural term for the space between two arches or the space between an arch and the rectangular frame around it. Decorative works of art have been inserted into these spaces, even though they were not designed for that purpose; they are merely by-products of architectural design. Gould and Lewontin 1979 have proposed that the term spandrel be used to denote biological features that are not directly the products of natural selection. The human chin might be an example.*

out in the previous chapter. Indeed, it may not be too fanciful to liken them to the distortions of the inside of a glove while a hand is formed into different handshapes. Of course, we do not see most of the movements, save those of the lips, when we watch people speaking. Even so, the perception of speech may depend at least in part on a sense of what the articulators are doing rather than on pure acoustic analysis, an idea expressed in the so-called "motor theory" of speech perception.[41] Even what we *see* the articulators doing can influence what we hear. If you dub a sound such as *ga* onto a video of a mouth that is actually saying *ba*, then you *hear* the syllable *da*, which is a sort of compromise between the sound itself and what the lips seem to be saying. This phenomenon is known as the McGurk effect.[42] Ventriloquists are also able to trick us into thinking the dummy is talking by keeping their own lips as still as possible, while moving the dummy's mouth in time with their own utterances. Even speech itself has not escaped from its gestural origins, it seems. Rather, it may have simply extended the gestural repertoire, or rendered audible gestures that are at least partly invisible.

The alterations of the vocal tract may not have been driven entirely by voicing. We can communicate in a reasonably articulate manner without using the voice at all, as when we whisper. Many speech sounds, such as /f/, /s/, /t/, /p/, /k/, and so on, are not voiced, and the impact of voicing on them is to create a new set—/v/, /z/, /d/, /b/, and /g/, respectively—thereby increasing the overall range of sounds. Oral gestures may have originated as unvoiced sounds, such as clicks or something akin to the lip smacking and teeth chattering of chimpanzees (described in more detail below), or even as visible signs, with voicing later added to extend the range and make hidden gestures accessible to the ear.

Although the lowering of the larynx was critical in the evolution of human speech, it may also have played another, and quite different, role. In general, the larger the animal, the longer the vocal tract, and formant frequencies depend in turn on the length of the vocal tract. Large animals with long vocal tracts tend to have low formants—in other words, deep voices. It is therefore possible that the lowering of the larynx in humans might have been selected to make us sound larger than we actually are, and so scare off potentially dangerous predators. Some birds, bless them, have evolved elongated

[41] *Liberman and Mattingly 1982.*
[42] *McGurk and MacDonald 1976.*

trachea that coil round inside the body, greatly lengthening the length of the vocal tract and lowering the formant frequencies, presumably enhancing the impression of size.[43] So the next time you are threatened by a lion, don't just scream; instead, roar, with the deepest voice you can muster.

These modifications to the vocal tract were not achieved without cost. The lowering of the larynx means that breathing and swallowing must share the same passage. People, unlike other mammals, cannot breathe and swallow at the same time and are therefore especially vulnerable to choking. If this was the price we paid for speech, then speech must have been of considerable adaptive significance in human evolution, although we might be warned by the words of Lucretius in *De Rerum Natura*: "From the midst of the fountain of delights rises something bitter that chokes them all amongst the flowers."

The larynx does not begin its slow descent until about three months of age (which is why babies can suckle and breathe at the same time), and it settles in its lower position after three or four years. A second, smaller descent takes place in males at puberty, causing the young man's voice to break—presumably so that his deeper voice will be more effective in scaring away predators. That's not the end of it, though. In Shakespeare's *As You Like It*, Jaques notes that late in life there comes a time for a chap when

> his big manly voice,
> Turning again towards childish treble, pipes
> And whistles in his sound.

It is not only the lowering of the larynx that sets us apart from the other great apes. Figure 7.2 shows that the chimpanzee, like the other great apes and many other primates, has a laryngeal air sac that extends out from the larynx beneath the skin of the neck and thorax. This can hold up to six liters of air, and almost certainly plays a role in vocalization, but it is totally absent in humans. Tecumseh Fitch speculates that it may be important in loud calls, but not in speech; again speech seems to be fundamentally different from animal vocalizations.

When in the evolution of our species did these remarkable changes in the human vocal tract occur? One clue comes from a consideration of the nerve, known as the hypoglossal nerve, that innervates the muscles of the tongue. In mammals, this nerve passes

[43] *See Fitch 2000 for further discussion of the so-called "size exaggeration" hypothesis.*

through the hypoglossal canal at the base of the skull. In humans the hypoglossal canal is much larger relative to the size of the oral cavity than it is in the great apes, presumably because the production of speech requires relatively more motor units than are needed for ape calls. Measurements from the skeletal remains of fossils also show that the hypoglossal canal in early australopithecines, and perhaps in *Homo habilis*, was about the same size as it is in modern great apes. In contrast, two Neanderthal skulls and one early *Homo sapiens* skull contained hypoglossal canals well within the modern human range. From this it has been concluded that the capacity for speech, or something like it, was present by at least 300,000 years ago, which is the approximate date of the earliest Neanderthal skull.[44] It is generally recognized that the Neanderthals were distinct from *Homo sapiens* but share a common ancestor dating from some 500,000 years ago,[45] and it might be reasonable to conclude that this common ancestor also possessed sufficient control of the tongue for articulate speech.

Philip Lieberman has long argued, though, that the changes that resulted in the modern human vocal tract were not complete until the emergence of our own species around 150,000 years ago.[46] In his view, they were also incomplete in the Neanderthal even as recently as 30,000 years ago. In human children, the lowering of the larynx in the first few years of life is accompanied by a flattening of the face, so that, relative to chimpanzees and other primates, we humans have mouths that are short from back to front.[47] The length of the mouth more or less matches the length of the pharynx, the other branch of the right-angled vocal tract. Lieberman argues that the two branches of the tube must be of approximately equal length to enable us to produce the point vowels that define the range of vowel sounds we use in normal speech. The fossil evidence shows that our Neanderthal cousins did not have flattened faces like ours; they had long mouths, more like those of apes. Since the flattening of the face had apparently not occurred in the Neanderthal, we can reasonably assume that the lowering of the larynx had also not taken place, or was at least incomplete.

Moreover, for the length of the Neanderthal pharynx to match the length of the Neanderthal mouth, the larynx would have to have been located in the chest. This would surely have prevented the poor creatures from swallowing! While

[44] Kay, Cartmill, and Barlow 1998.
[45] Ward and Stringer 1997.
[46] The very title of his 1998 book is *Eve Spoke. Eve*, whom we met in chapter 3, is the 150,000-year-old woman said to be the source of present-day variation in mitochondrial DNA. See also Lieberman, Crelin, and Klatt 1972.
[47] D. Lieberman 1998.

this might explain why the Neanderthals became extinct, it is more plausible to suppose that the changes to the face and vocal tract that have given us the power of articulate speech had not yet occurred, or were incomplete, in the Neanderthal.[48] If Lieberman's argument is correct, the fully formed human vocal tract must have emerged *since* the parting of the ways between *Homo sapiens* and the Neanderthals.[49] Indeed, it might be considered a critical part of the "speciation event" that gave rise to our own species some 150,000 years ago.

But it is unlikely that speech itself arrived suddenly. Even Lieberman has acknowledged that the Neanderthals could speak, but without the full range of articulation possessed by *Homo sapiens*. Perhaps they had the vocal range of modern human infants. The alterations to the vocal tract, along with other changes (to be reviewed below), must surely have occurred gradually, perhaps reaching their present level of elaboration with the emergence of our species.

■ A breath of air ■

Speaking requires precise control over breathing. We speak as we breathe out, on the outgoing tide as it were, so that in uttering a long sentence, or making a speech, we breathe out slowly, every now and then taking sharp intakes of breath to renew the air supply. Whew. If you try to speak as you breathe in, you will find that the articulators work well enough, but the actual voicing is reduced to an unpleasant croak. Perhaps this is where frogs went wrong. During speech, the outward flow of breath must be precisely controlled to provide the intonations and emphases of normal speech. Although relatively little is known about how nonhuman primates breathe while vocalizing, it seems that they produce sounds on both the intake and outflow of breath, and that only one sound unit is produced in each individual movement of air. By contrast, a single outward breath during human speech normally ranges from about two to six seconds and can be more than twelve seconds,[50] and

[48] These points are reiterated by Philip Lieberman in his recent book, Human Language and Our Reptilian Brain 2000.

[49] Lieberman's argument is based in part on hypothetical reconstructions of the vocal tract from fossil evidence and has proven highly contentious; see, for example, Gibson and Jessee 1999. The vowels are not all that critical to speech, and we can make ourselves reasonably well understood when we speak through clenched teeth, or wittingly eliminate all variation, *f y** *nd*rst*nd wh*t * m**n. Falk 1975 has argued that if Lieberman's reconstructions of the Neanderthal vocal tract were correct, the poor creatures would have been unable to swallow. Lieberman 1982 has retorted in turn that if Falk were correct, the chimpanzee would also be unable to swallow. So it goes.

[50] Winkworth, Davis, Adams, and Ellis 1995.

of course it includes a great variety of different sound units. I bet you can easily utter this sentence on a single breath.

Of course, the main reason why animals breathe is to supply air to the lungs and so provide the oxygen we need to keep us alive. The role of breathing in speech is another example of exaptation. Nevertheless, there are differences between the way we breathe for air, known as "quiet breathing," and the way we breathe to produce speech. For speech, the respiratory muscles keep the air pressure just below the larynx—the subglottal air pressure—more or less constant, so that the voice can be modulated regardless of how full the lungs are. You can even empty the lungs and keep on talking for a bit, albeit with a little strain. (Go on, try.) This control is quite complex, since different patterns of muscle activity are required, depending on the volume of air in the lungs.[51]

Quiet breathing involves movements of the diaphragm, whereas the more precisely controlled breathing that is required for speech requires extra muscles in the thorax and abdomen. The muscles of the diaphragm involved in quiet breathing are innervated by the vagus nerve, but the muscles of the thorax and abdomen that are involved in speech are innervated through the thoracic region of the spinal cord. In modern humans this region is considerably larger than it is in nonhuman primates, including the great apes, presumably reflecting the extra demands placed on these muscles by speech. Studies of hominin fossils show that this enlargement was not present in the early hominins, or even in *Homo erectus*, who dates from about 1.9 million years ago. However, it was clearly present in several Neanderthal fossils.[52] At least a rudimentary form of speech may, therefore, have evolved in the common ancestor of ourselves and the Neanderthals by some 500,000 years ago.

These adaptations in the control of breathing, along with the modifications of the vocal tract, are as important for singing as for speech, if not more so. Charles Darwin, seduced by birdsong, thought that speech might have evolved from singing, largely because it was sexy:

> When we treat of sexual selection we shall see that primeval man, or rather some early progenitor of man, probably first used his voice in producing true musical cadences, that is in singing, as do some of the gibbon-apes at the present day;

[51] Mead, Bouhuys, and Proctor 1968.
[52] MacLarnon and Hewitt 1999.

and we may conclude from a widely-spread analogy, that this power would have been especially exerted during the courtship of the sexes,—would have expressed various emotions, such as love, jealousy, triumph,—and would have served as a challenge to rivals. It is, therefore, probable that the imitation of musical cries by articulate sounds may have given rise to words expressing of various complex emotions.[53]

Others have pursued a similar theme.[54] We have already seen that rhythmic drumming and pant hoots in chimpanzees seem to provide a way of maintaining bonds between members of a troop, and song-like duetting is also common among other Old World primates, including indris, titis, tarsiers, and gibbons.[55] Oh, and birds sing, do they not? —and probably for much the same reasons. Music is about as ubiquitous a human activity as speech, although most of us are a good deal better at speaking than we are at singing. Moreover, propositional speech, although not always unambiguous, conveys information that is much more precise in its meaning and reference than music, which tends to convey vague emotions or holistic impressions rather than exact meaning—although I do not wish to belittle the complexity of a Beethoven symphony. People are more likely to disagree over the message contained in a piece of music, no matter how carefully crafted, than over that contained in a meticulously worded legal document. One aspect of speech that may well derive from song is *prosody*, the variations in intonation and loudness that signal different kinds of sentences: a barked order has a different quality from a timid question, and you can probably tell the difference even in a language you are not familiar with. To suppose that early hominins went through a phase of universal singing, presumably round the campfire, or yodeling across the valleys, is to go well beyond the evidence,[56] but perhaps it has a ring, as it were, of plausibility.

■ Changes in the brain ■

[53] *Darwin 1896, 87.*
[54] *Richman 1993; Vaneechoutte and Skoyles 1998.*
[55] *Haimoff 1986.*
[56] *Going beyond the evidence to postulate earlier phases in which communication was carried out in a different way is not in general a good thing to do. Present company excepted, of course.*

Changes in the vocal tract, in the innervation of the tongue, and in the control of breathing were probably all necessary for the emergence of speech, but they were not sufficient. To make speech sounds, we must precisely synchronize sound production with movements of the articulators, such as the tongue and lips, and to actually speak, we must have access to the brain structures that govern perception and knowledge of the world, and that determine the things we want to talk about. To put it all together requires complex programming, and for that we need a brain—or at least half a brain. We use about one hundred muscles when we speak and produce phonemes at the rate of about ten to fifteen per second.[57] Given the variety of sounds that we cobble together when we speak, it is highly unlikely that this can be done without at least some involvement of the cerebral cortex, that most recently evolved, wrinkled surface of gray matter that is such a prominent feature of the human brain.

First of all, of course, the brain simply got bigger. As we saw in chapter 5, the human brain is about three times as big as one would expect for a primate of our size. The increase is especially marked for the cerebral cortex and for the ratio of gray matter to white matter[58]—vindicating Hercule Poirot's constant references to the importance of his "little gray cells."[59] More than that, the frontal lobes, important for short-term memory and planning, are disproportionately enlarged, and that development must surely bear on the production and understanding of long sentences like this one.[60] In addition, the frontal lobes are more wrinkled, or "gyrified," than other parts of the human brain.[61] This wrinkling is the brain's way of cramming a larger surface into a small space, as one might crinkle a sheet of paper to fit it into a small box.

But the brain changed in other ways as well. Noam Chomsky once remarked, "a chimpanzee is very smart and has all kinds of sensorimotor constructions (causality, representational functions, semiotic functions, and so forth), but one thing is missing: That little part of the left hemisphere that is responsible for the very specific functions of human language."[62] Chomsky may have exaggerated the intellectual prowess of the chimpanzee, but the "little part" that he thought missing from the chimpanzee is presumably Broca's area, which cropped up briefly in chapter 3. This area lies in the left frontal cortex, just in front of the area that controls movements of the mouth and of the

[57] For a succinct account of how we produce speech, from intention to actual production, see Levelt 2000.

[58] Rilling and Insel 1999.

[59] Hercule Poirot is the Belgian detective who features in many of the crime stories written by Agatha Christie.

[60] See Deacon 1997.

[61] Rilling and Insel 1999.

[62] Quoted in Piattelli-Palmarini 1980, 182.

hands. It is named after Paul Broca, the young French physician who discovered in the 1860s that damage to this area resulted in the loss of articulate speech. One of Broca's patients was known as Tan, because *tan* was the only articulate sound he could make. He was nevertheless able to understand the speech of others and could move his lips and tongue according to command. It appears that Broca's area is critically involved in the organization of spoken language, although it seems to play this role only in humans. Damage to the area corresponding to Broca's area in the monkey, along with damage to surrounding areas and to the corresponding areas on the right side of the brain, does not have any discernible effect on the animal's vocalization.[63]

Shortly after the discovery of Broca's area, the German neurologist Carl Wernicke found that damage to a posterior portion of the brain, around the junction of the left temporal, parietal, and occipital lobes of the brain, results in loss of comprehension of speech. These and other observations have clearly established that the left cerebral cortex is critical to language, at least in the great majority of people. There is also evidence that Wernicke's area, like the frontal lobes, is more gyrified than other parts of the brain—wrinkles bring wisdom in more ways than one.[64] For a long time it was thought that language was essentially a matter of Wernicke's area comprehending language and Broca's area producing it, with interconnecting fibers to make sure that what we say makes sense. This simple view is almost certainly wrong. Damage to Broca's area results in some loss of the ability to unravel the syntax of *heard* sentences, so Broca's area may not be involved only in speech production. It may have a more general role in syntax.

Patterns of imprints on the insides of fossil skulls suggest that these changes to the brain may have begun to develop with *Homo habilis*, some 2 million years ago. These imprints can be revealed on the surface of endocasts made by filling the insides of skulls or skull fragments with latex, a technique pioneered by Ralph L. Holloway. Endocasts show where the fissures, or grooves, are located on the surface of the brain, so that the sizes and locations of various lobes and other distinctive regions can be estimated. There appears to be an enlargement, corresponding to Broca's area, in the left

[63] Jürgens, Kirzinger, and von Cramon 1982.
[64] But at the expense of beauty, perhaps. As Thomas Nashe (1567–1601) put it:
Beauty is but a flower
Which wrinkles will devour.
—Summer's Last Will and Testament (1600)

side of the brain of *H. habilis*.[65] There is also evidence that the so-called inferior parietal lobule in *H. habilis* is larger than it is in the chimpanzee or in the australopithecines, although some dispute this.[66] If real, this enlargement can be attributed at least in part to an expansion of Wernicke's area. Philip V. Tobias was sufficiently impressed with these claims to declare: "The occurrence of both a strong inferior parietal lobule and a prominent motor speech area of Broca in the endocasts of *H. habilis* represents the first time in the history of early [hominins] that the two most important neural bases for language appear in the paleoneurological record."[67] Tobias's pronouncement must be qualified, however, in the light of a recent claim that the temporal planum, an area that overlaps with Wernicke's area, is larger on the left than on the right in the chimpanzee.[68] This is further discussed in the next chapter.

All of these changes in the brain, however, may have to do with *language* rather than speech, as has become clear from studies of those whose native language is signed. Like spoken language, signed language is impaired following damage to the left side of the brain, and in much the same ways.[69] Brain-imaging studies have shown that Broca's area is activated when deaf signers produce either words[70] or sentences[71] in ASL. Another recent study by Helen Neville and her colleagues, using positron emission tomography (PET) to reveal brain activation, has nevertheless provoked controversy, because it shows extensive activation of the right side as well as the left when deaf people viewed a video of a signer producing sentences in ASL.[72] Even so, the areas activated in the left side of the brain were essentially the same as the areas activated in hearing people when they read printed English. A number of studies have shown that the right side of the brain is also activated when people listen to spoken language.[73] The rather more diffuse activation of the right brain in the deaf video watchers may simply relate to the more spatial aspect of signed language.[74] But in any event it seems that the *production* of signs activates only the left brain, con-

[65] *Holloway 1983.*

[66] *Holloway 1985 maintained that the enlargement of the parietal lobe was already under way in the australopithecines, while Falk 1983 has argued that it was only apparent in H. habilis.*

[67] *Tobias 1987, 753.*

[68] *Gannon, Holloway, Broadfield, and Braun 1998.*

[69] *Kimura 1981; Poizner, Klima and Bellugi 1987.*

[70] *Hickok, Bellugi and Klima 1997 used functional magnetic resonance imaging (fMRI) to show this.*

[71] *A study by McGuire, Robertson, Thacker, David, Kitson, Frackiowak, and Frith 1997, using positron emission tomography (PET).*

[72] *Neville, Bavelier, Corina, Rauschecker, Karni, Lalwani, Braun, Clark, Jezzard, and Turner 1997.*

[73] *See Corina, Neville, and Bavelier 1998 for a summary.*

[74] *For a more in-depth discussion and interpretation of the results obtained by Neville and her colleages, see the exchange between Hickok, Bellugi, and Klima 1998 and Corina, Neville, and Bavelier 1998.*

sistent with the view that syntax, the sine qua non of true language, is in most of us the property of the left hemisphere.

Further, we saw in chapter 3 that an area in the monkey brain corresponding to Broca's area is the site of the "mirror neurons" that map perceived reaching and grasping movements onto those performed by the animal itself. An area in the left parietal lobe of the human brain, close to Wernicke's area and possibly overlapping it, seems to store programs for executing skilled actions, including *manual* actions.[75] Indeed, this area seems to be part of the human mirror-neuron system discussed in chapter 3, since damage to it may cause an impairment in the ability to recognize skilled actions, as well as in the ability to perform them.[76] My guess is that the enlargements and wrinkling of these language-related parts of the brain had to do with language itself, and perhaps more generally with the planning and execution of complex action, and not specifically with speech. They may have led to the emergence of syntax, but initially, at least, in the context of gestural communication rather than speech.

But speech, as it gradually emerged, created its own demands on the brain. As we have seen, even the breath control required for speech is complex, and, not surprisingly, research reveals that it may be at least partly under the control of the cerebral cortex. The famous neurosurgeon Wilder Penfield showed in the 1930s that stimulating the so-called motor cortex of the brain sometimes caused patients to vocalize, usually without moving their lips or tongues. This vocalization, though inarticulate, was speechlike in that it occurred only when the patients breathed out.[77] A more recent study has shown that activity in the same area of the motor cortex, along with activity in the sensorimotor cortex and the supplementary motor area in the frontal lobes, was associated specifically with breathing during speech.[78] It has also been shown that the motor cortex is active when people breathe out willfully without vocalizing.[79] It is interesting that the control of breathing in these studies seems to have involved both sides of the brain equally, whereas the articulate aspects of speech are controlled by the left side of the brain in most people.

We now know that subcortical regions, especially the so-called basal ganglia, are also critically involved in speech.[80] It would therefore be wrong to claim that the control of speech is entirely dif-

[75] Heilman and Rothi 1985.
[76] Heilman, Rothi, and Valentine 1982.
[77] Penfield and Bouldrey 1937.
[78] Murphy, Corfield, Guz, Wise, Harrison, and Adams 1997.
[79] Ramsay, Adams, Murphy, Corfield, Grootoonk, Bailey, Frackowiak, and Guz 1993.
[80] See Lieberman 2000 for detailed discussion.

ferent from the control of vocalization in nonhuman primates. As Darwin himself put it, evolution is a matter of "descent with modification":[81] some of the mechanisms of vocalization were extended in humans to include cortical structures, and we thereby acquired a much greater degree of flexibility and sequential organization than was possible in our primate forebears.

■ Chewing the fat ■

Another perspective on the precursors to speech is provided by Peter MacNeilage, who has argued that speech evolved from "ingestion-related cyclicities of mandibular oscillation,"[82] which is itself, I suppose, a bit of a mouthful. Repeated movements of the mouth and jaw have been around as long as animals have chewed their food. Human speech resembles eating in that it consists of repeated alternations of opening and closing the vocal tract, as is assiduously practiced by babies when they start babbling. Speech is then built on a modulation of this repetitive pattern. Indeed, one can observe the transformation taking place as young babies develop from repeating the same syllables, as in *ba-ba-ba* or *ga-ga-ga*, to producing variations, as in *ba-bee* or *da-dee*. This is followed, in the second year of life, by the rapid accumulation of words, which are later combined to form sentences. In these respects, the development of sound sequences during infancy and early childhood provides a fossil record, as MacNeilage puts it, of true speech. It is at least the nearest thing to a fossil record we have, since speech itself does not fossilize.

Animal calls tell us little about the evolution of speech itself, since they tend to be holistic, and if repeated, do not involve different arrangements of the subcomponents, as words do. Perhaps a better clue comes from nonvocal movements of the mouth. Nonhuman primates make use of repetitive movements like lip smacking, tongue smacks, and teeth chatters as means of communicating. The most common of these is the lip smack, in which the lower jaw moves up and down, the lips open and close a little, and the tongue moves forward and back between the teeth. ("Lip smack" is something of a misnomer, since it is really the tongue that does the work, although this is difficult to detect visually.) Lip smacking is usually audible but does not involve voicing. It tends to be used in one-to-

[81] *Darwin 1859, 240.*
[82] *MacNeilage 1998, 499.*

one social interactions and sometimes the participants take turns at it.[83]

MacNeilage suggests that these repetitive sequences, as well as the sequencing of human language, are built on the repetitive opening and closing of the jaw that occurs in the ingestion of food. Food for talk, as it were. It would indeed be surprising if speech did not somehow involve the mechanisms of chin wagging—or mandibular oscillation, if you like long words—which probably goes back to our vertebrate origins. And eating is not the only cyclical action we indulge in; we are a-tremble with rhythms of one sort or another, including breathing, sucking, walking, swimming, and even, dare I mention it, sex. It is likely that these different cyclical activities share a common rhythm generator in the brain stem but have been progressively modulated to perform different functions by higher centers in the brain.

The rhythm of speech forms what MacNeilage calls the *frame*, into which the *content* must be inserted—much as one fills a diary with words. The content is essentially supplied through the cortical areas, including Broca's and Wernicke's, which provide access to information about the external world. MacNeilage suggests that the meeting point between frame and content is indeed Broca's area, citing evidence that this region is part of a network that has to do with mastication in primates.[84] But whether Broca's area is truly a center for mastication is controversial;[85] another part of the frontal cortex, the supplementary motor area, may play a more important role than Broca's in speech production.[86] Perhaps the role of Broca's area in mastication has more to do with its general role in the programming of actions, be they manual or oral.[87]

MacNeilage's proposals are controversial, but Giacomo Rizzolatti, co-discoverer of the "mirror neurons," offers a helping hand. Discussing MacNeilage's theory, he argues that the importance of Broca's area in the evolution of speech is not that it was involved in movements of the mouth and face but rather that it served to map internally generated hand movements onto perceived hand movements made by others.[88] Language, he as-

[83] Redican 1975.

[84] One cannot but admire the versatility supposed of Broca's area. Descartes argued that the pineal gland was the seat of the soul, but Broca's area may have a stronger claim.

[85] See the commentaries that immediately follow MacNeilage's 1998 article, especially those by Abbs and DePaul; Abry, Boë, Laboissière, and Schwartz; and Sessle.

[86] Abry et al. 1998.

[87] It might be interesting, in fact, to look for neurons in the homologue of Broca's area in the monkey that respond to movements of the mouth in the way that neurons there respond to reaching and grasping with the hands. After all, the mouth is also sometimes useful for grasping, often serving as a third hand.

[88] Rizzolatti 1998.

serts, was born in manual gestures—and who am I to disagree? Movements of the mouth and vocalizations were later co-opted, perhaps in part because primates use gestures of the mouth along with manual gestures as forms of intentional communication. Language proceeded not just from hand to mouth affair but also from mouth to voice. This scenario explains why early species of *Homo*, such as *Homo habilis*, had a well-developed Broca's area but a primitive vocal tract that would not have been able to sustain normal speech.

■ Objections to gestural theory ■

Of course, not all experts agree that language evolved from gesture; indeed, it is probably still a minority view. This may be the appropriate point to examine some of the objections.

John Bradshaw, in his 1997 book *Human Evolution: A Neuropsychological Perspective*, writes, "[An] argument against a gestural stage in language evolution is the probability that increasing tool behavior would have hampered gesture and so have tended to favor speech." But then he goes on to say, "though one can of course argue that gestures might have held back the development of tool behaviors until after language switched to the oral/acoustic mode."[89] That is what I am indeed arguing in this book, although I prefer to say that tool behavior was not so much held back by gesture as permitted to flourish when the hands were freed from linguistic duty.

In his entertaining and popular book *The Language Instinct*, Steven Pinker writes:

> Sign language has frequently been suggested as an intermediate, but that was before scientists discovered that sign language was every bit as complex as speech. Also, signing seems to depend on Broca's and Wernicke's areas, which are in close proximity to vocal and auditory areas in the in the cortex, respectively. To the extent that brain areas for abstract computation are placed near the centers that process their inputs and outputs, this would suggest that speech is more basic.[90]

Yet we have seen, in chapter 3, that the homologue of Broca's area in the macaque is the home of the "mirror neurons" that have to do with

[89] *Bradshaw 1997, 102.*
[90] *Pinker 1994, 352.*

the production and perception of gestures, not of vocalization, and even in humans Broca's area responds to the visual perception of gestures. Wernicke's area straddles parietal, occipital, and temporal lobes—a multinational zone sometimes referred to as the POT[91]—and might therefore be considered no more auditory than visual, since the occipital cortex is primarily involved in visual analysis. Indeed, it seems ideally located for the analysis of both vocal and manual action, and for the integration of the two.

Pinker goes on to suggest that the alarm calls of vervet monkeys, which I discussed in chapter 2, might offer a better clue as to the origins of language, partly on the grounds that they are "quasi-referential." Maybe, in hominins, such calls came under voluntary control and were eventually produced in sequences to create more complex meanings. I think, though, that manual gestures provide a better basics on which to build sequences of voluntary actions, at least in primates, and probably in early hominins. But Pinker admits that his speculations may have no more in their favor than Lily Tomlin's idea that the first human sentence was "What a hairy back!" Mine too, perhaps.

In writing this book I have been indebted to Terrence Deacon, whose 1997 book *The Symbolic Species* eloquently supplies much of the evidence that primates are much more proficient manually than vocally, at least with respect to voluntary control and sheer flexibility of action. Yet Deacon also stops short of endorsing the gestural theory. His point is a subtle one, and worth quoting at some length:

> If something analogous to American Sign Language long predated spoken language and served as the bridge linking the communication processes of our relatively inarticulate early ancestors, then we should expect that a considerable period of Baldwinian evolution would have specialized both the production and the perception of manual gestures. Clearly, there are some nearly universal gestures associated with pointing, begging, threatening, and so on, but these more closely resemble the nonlinguistic gestural communications of other primates both in their indexical functions and in the sorts of social relationships they encode, rather than anything linguistic or symbolic. The

[91] *Wilkins and Wakefield 1995.*

absence of other similarly categorical and modularized gestural predispositions suggests, in comparison to speech specializations, that the vast majority of Baldwinian evolution has taken place with respect to speech.[92]

There are a number of points to observe here. First, the gestures of "pointing, begging, threatening, and so on" are probably more languagelike than animal calls, as I explained in chapter 3, not least because they are under voluntary control. Second, we also saw that primate gestures tend to become conventionalized, and therefore increasingly symbolic rather than iconic. Third, Deacon may well be correct in suggesting that the shaping of the speech organs may have occurred through Baldwinian evolution, but the selective pressure may not have been toward language, as Deacon seems to imply, but rather toward speech itself. There was in fact little need for evolution to shape "the production and perception of manual gestures," since these were already largely preadapted for effective communication— and remember those handy mirror neurons. It was a bit like inheriting grandma's old car, which was perfectly fine until we could afford a new one.

■ Getting it together ■

I hope I have at least made it clear that a lot of things had to happen in the descent from ape to human before we could speak. Speech was almost certainly impossible for the common ancestor of ourselves and the chimpanzee, and there was probably little substantial change prior to the appearance of *Homo*, that is to say, for the next 3 or 4 million years. The emergence of articulate speech since that time almost seems a miracle, involving wholesale changes to the vocal tract, the control of breathing, and the brain.

But we must always be wary of miracles. The transition appears somewhat less miraculous if we suppose that it was built on a scaffolding of manual communication that had deep roots in primate

[92] *Deacon 1997, 362. Baldwinian evolution refers to behavioral effects on natural selection. A concept proposed by the American psychologist James Mark Baldwin, it provides a way of avoiding the view that behavior itself can lead to evolutionary change, generally regarded as a Lamarckian heresy. Instead, behavior itself can change the environment, and the new environment then creates new selective pressures. For example, most mammals become intolerant to the milk sugar lactose after weaning. It is not entirely clear to me, however, why one need appeal specifically to Baldwinian evolution to account for the selective pressures that might have led to more sophisticated language. Those pressures may have arisen simply from natural changes in ecological conditions or from changes in behavior itself—or from a combination of the two. The point is not especially critical, however, since it is highly likely that selective pressures favored the evolution of a more sophisticated language, whatever the underlying mechanisms.*

evolution. Within hominin evolution, the upright stance may have enhanced gestural communication, perhaps leading to greater elaboration of gestural signals and the beginnings of mime, although it is unlikely that anything resembling syntax developed until the emergence of the larger-brained genus *Homo* some 2 million years ago. The australopithecines may have developed communication skills to the level of what Bickerton has described as protolanguage, but probably no further.

Speech was probably also impossible for the earliest members of the genus *Homo*. Yet the emergence of this genus signals a number of behavioral and morphological changes strongly suggestive of *language*, if not of speech. As we saw in chapter 5, this was when brain size began to markedly increase, and at roughly the same time stone tools begin to appear in the archeological record. Migrations out of Africa can be dated from nearly 2 million years ago. More tellingly, perhaps, the fossil evidence for language production areas on the left side of the brain also appears to date from this period. All of this may coincide with the beginnings of syntax, which brings with it a more generative form of communication—something beyond protolanguage. But this language would have to have been gestural rather than vocal, since the changes in the vocal tract and the control of breathing necessary for articulate speech were yet to come.

This is not to say that vocalization played no part in the language of early *Homo*. At first, grunts and cries, not to mention pant hoots, may have punctuated gestural communication, adding emphasis and emotional tone. Arthur Sigismund Diamond, an English comparative linguist, suggested that speech may have originated in the release of air that follows action, as in the grunting of many of today's tennis players when they play a shot. The muscles involved in moving one's arms are attached to the ribs, and if the ribs are to provide a firm base from which to contract the arm muscles, it is often necessary to hold one's breath. As we saw earlier, the larynx is involved in closing off the lungs when one hold's one breath, so the sudden release of air causes a grunt. Diamond also suggested that early vocalizations accompanied violent actions, such as cutting, breaking, hammering, smashing, and so forth, and that early language was essentially a mimicking of such actions—or "suggestions of action," as Diamond put it.[93]

---

[93] *A. Diamond 1959.*

It may well be the grunt that provides the basis of the syllabic structure of speech. As noted earlier, most speech consists of repetitions of consonant-vowel syllables, beginning with babbling and later developing into words. The syllable *ba*, for example, is essentially a modified grunt, with the release of air controlled by the lips rather than the larynx. Eventually, then, the grunts may have become so articulated as to provide the essential core of language itself, with manual gestures playing an ever more subsidiary role.

Precisely when speech would have assumed dominance is largely a matter of conjecture, and the Neanderthals may well provide the key. As we have seen, many of the adaptations necessary for articulate speech seem to have been present in the Neanderthals, including enriched innervation of the tongue and the enlargement of the thoracic canal necessary for the control of breathing in speech. Yet Lieberman has argued that the vocal tract had still not assumed the shape necessary for making the point vowels. The Neanderthals had brains that were, if anything, slightly larger than our own, which suggests that they may have been no less intelligent and no less articulate. Our own species, *Homo sapiens*, inhabited the same regions of Europe up until about forty to thirty thousand years ago, when the Neanderthals died out, yet the two species apparently did not interbreed.

So what kept them apart, and why did *sapiens* prevail? As we have seen, Neanderthals did not have the flattened face that characterizes our own species, so they may have simply seemed gross and ugly to our refined forebears. Neanderthal skulls were first discovered in the Neander Valley in Germany in 1856, and as John E. Pfeiffer put it, the Neanderthal "came into the world of the Victorians like a naked savage into a ladies' sewing circle."[94] The sewing circle of thirty-five thousand years ago might have reacted similarly. But a further impediment to social intercourse may have been language. The conversion to autonomous speech may have been incomplete in the Neanderthals, making them seem inarticulate to the good ladies of the *sapiens* species; further, they might have been forced to rely more on manual gestures than on speech. Indeed, as I suggested above, even early *Homo sapiens* may have relied at least partly on manual gestures, perhaps because the possibility of autonomous speech had not yet occurred to them. But, along with the Neanderthals and the rem-

[94] *Pfeiffer 1973, 160.*

nants of *Homo erectus*, they were essentially—to put it euphemisti-cally—talked out of existence by our chattering forebears, who mi-grated out of Africa around fifty thousand years ago.

Why was speech so critical, leading to the evolution of such com-plex changes to the mouth and throat that we forever run the risk of choking? If articulate language can be accomplished solely through gesture, why did the relentless selective mechanisms of evolution force us to replace it with a system that was initially much more prim-itive? I will attempt to answer this question in chapter 9. But first I want to discuss another pointer to the hand-to-mouth progression of language, one that is dear to my lopsided heart.

8 ■ Why Are We Lopsided? ■

As I have several times had occasion to mention, language is largely a function of the left side of the brain, at least in the great majority of us. Broca's and Wernicke's areas, the brain territories classically associated with articulate language, are left-hemispheric, as are the areas critical to reading, writing, and signed language. It is also true, of course, that the great majority of us are right-handed. Since the right hand is controlled largely by the left side of the brain, our left-lobed lopsidedness seems to define another link between hand and mouth in human evolution—perhaps a further pointer to the gestural origins of language. Could it reflect the unique blending of vocal and gestural language that seems to characterize our species?

If that were so, then we might also expect cerebral asymmetry itself to be a feature of humanity, integrating several characteristics that appear to be uniquely human.[1] As I have explained, true language seems to be restricted to our own species, unless there are extraterrestrials out there who have discovered syntax. Our right-handedness is also linked to the fact that we are, as a species, exceptionally skilled in manual activities—not for nothing is manual skill known as *dexterity*. What other species can thread a needle, throw a ball with such extraordinary accuracy, build a house, or write a

[1] *It was the principal theme of my 1991 book* The Lopsided Ape. *As the title of this chapter indicates, I have not completely reformed.*

thank-you letter to a generous aunt[2]—in most cases with the right hand playing the major role? Some authors have sought to link even deeper human attributes with the left hemisphere of the brain. For example, the eminent physiologist Sir John C. Eccles, echoing Descartes, argued that the right hemisphere is a mere "computer," comparable to the brains of lower animals, while the left hemisphere endows us with free will and self-consciousness.[3] Although Oliver Zangwill described this as "little more than a desperate rearguard action to save the existence and indivisibility of the soul,"[4] the idea is curiously persistent.

Julian Jaynes, in his provocative book *The Origins of Consciousness in the Breakdown of the Bicameral Mind*, argued that until about three thousand years ago humans were guided in their actions and decisions by hallucinations, generated by a "bicameral" brain and interpreted as the voices of the Gods.[5] According to Jaynes, cerebral lateralization arose in response to disasters that occurred in the second millennium B.C., including floods, earthquakes, mass migrations, conquests, and stock market crashes, leading to the emergence of self-consciousness and individual responsibility for action, mediated by the left hemisphere. After that, people did not wait for the Gods to tell them what to do but instead decided for themselves. Jaynes argued that the change can be discerned in the different styles of the *Iliad*, which contains virtually no references to the self or first-person constructions, and the *Odyssey*, which does incorporate the first person and is more "modern" in tone.[6] Cerebral asymmetry, according to Jaynes, had nothing to do with language per se; language had evolved well before these momentous events took place, he says.

Jaynes's theory makes little sense in the light of evolutionary evidence already discussed in earlier chapters of this book. Evidence for asymmetry of brain areas related to language goes back at least 2 million years, and it is in any case unlikely that cerebral asymmetry could have evolved in the course of a single millennium. Nevertheless the idea that left-hemisphere specialization might have something to do with the sense of responsibility for action is one that has persisted. Michael Gazzaniga has concluded, on the basis of some

[2] A slightly unfair example, it must be admitted, since writing also involves language. Besides, not all of us are lucky enough to have generous aunts.

[3] Eccles 1965, 1981.

[4] Zangwill 1976, 304.

[5] Jaynes 1976.

[6] As every schoolchild knows, or once knew, the Iliad and the Odyssey were both attributed to Homer, and you may well protest that consciousness is scarcely likely to have vanquished the gods within the working life of one man. As Jaynes points out, though, we do not really know that such a person as Homer actually existed, and the events described in the Iliad probably date from about 1230 B.C., while those in the Odyssey date from between 1000 and 800 B.C.

thirty-five years of research with "split-brained" people, that is, people who have had the cortical connections between the cerebral hemispheres surgically cut for the relief of intractable epilepsy, that the left hemisphere functions as a general "interpreter," giving rise to "the feeling that we are in charge of our actions."[7] There is even an experiment involving the measurement of brain activity in normal people showing that the left brain is dominant in selecting which of two fingers to move in making a choice between visually presented shapes, regardless of which *hand* is used.[8]

Bu we must tread carefully here, since asymmetries, including cerebral asymmetries, are not themselves unique to humans, and it may be arrogant to think that only humans are capable of self-consciousness. Furthermore, not all humans conform to the general pattern of left-cerebral dominance for language and hand control: some individuals insist on using their left hands, or their right brains, in a carefree reversal of the norm. Let's therefore take a more general look at asymmetries in biological systems and then consider what, if anything, might be special about handedness and cerebral asymmetry in humans.

■ On symmetry and asymmetry ■

Our asymmetries gain significance only when we consider that we also display a striking bilateral *symmetry*, as do nearly all other animals. One might be tempted to think that symmetry is a sort of default condition that occurs in the absence of any selective pressure toward asymmetry, but this is clearly not so. We are, after all, constructed of molecules that are famously asymmetrical, as explicitly shown when the structure of the DNA molecule was revealed.[9] We are certainly more bilaterally symmetrical than one would expect by chance; it is as though our Maker has gone to special lengths to ensure that the left sides of our bodies are near-perfect mirror images of our right sides. It is also obvious that animals are not symmetrical with respect to top and bottom or back and front, so there must be special reasons why the left-right axis has been singled out for special treatment.

The asymmetry between top and bottom has to do basically with the influence of gravity. Selective pressures dictated that footlike structures should evolve on the part of the body that makes

[7] *Gazzaniga 2000, 1293.*

[8] *Schluter, Krams, Rushworth, and Passingham 2001. Brain activity was measured using positron emission tomography (PET).*

[9] *It has not escaped my notice that this was accomplished by Watson and Crick 1953.*

connection with the ground—something that even our friends the birds occasionally do. It is also adaptive to have eyes placed somewhat high up on the body, so as to gain better vision across terrain, as meerkats do when they stand tall to watch for predators. Of course, for some animals it's also not a bad idea to be able to bring one's eyes down to ground level, especially when foraging for food there—or looking for a dropped contact lens. Animals may therefore be equipped with flexible necks, so they can raise and lower the head and eyes, or flexible limbs, so they can drop to a crawling posture. One might also suppose that it's not a bad idea to have the orifice for the intake of food somewhat removed from the orifice for its excretion, and evolution kindly took care of that for us.[10] In other words, we can see a number of reasons why different organs are placed where they are on the top-bottom axis, and certainly no obvious pressures for *symmetry* along this axis.

The front-back axis became important once organisms evolved the capacity for movement. Trees and plants typically have no obvious back or front, being rooted to the ground. They're just not going anywhere, with the exception perhaps of the forest that moved in *Macbeth*. A messenger brings the news:

> Messenger: Gracious my lord,
>> I should report that which I say I saw,
>> But know not how to do it.
> Macbeth: Well, say, sir.
> Messenger: As I did stand my watch upon the hill,
>> I looked toward Birnam, and, anon, methought,
>> The wood began to move.
> Macbeth: Liar and slave! [striking him.][11]

The unfortunate messenger was right, and the wood continued its inexorable march toward Dunsinane, where the fretful Macbeth awaited his fate.

But this event was fairly exceptional even in literature, as far as I know. When other organisms evolved ways of moving about, selective pressure would have led to the evolution of systematic differences along the axis of the motion itself. Fronts became different from backs. Legs are shaped so that you can run much faster forward than backward, and so can your

[10] A friend who shall be nameless nevertheless wondered whether it was such a good idea to have the nose, which is apt on occasion to drip, quite so close to the mouth.

[11] Shakespeare's Macbeth, act 5, scene 5.

dog. Faces are in front of the head so you can smile nicely at people as you approach them—or bare your teeth in a snarl, if you want to scare them away. The eyes are swiveled to the front so you can see where you are going, although in many species, such as the horse, they are placed somewhat to the side, with some frontal overlap. Most birds also have eyes that are placed somewhat laterally, with owls a conspicuous exception.

Once the top-down and front-back axes have been defined, there is virtually no environmental pressure to distinguish the left and right sides. For an animal that can move freely about the earth's surface, the environment to its left is not systematically different from the environment to its right. Predators and prey are as likely to lurk on one side as on the other. Subtle effects, such as the Coriolis force that influences weather patterns and the way the water spirals as it goes down the waste hole in the bath, are so slight as to have no effect on the construction of the body. Still, even in the absence of systematic left-right differences in environmental influences, one might expect random asymmetries between the left and right sides of our bodies, and indeed there are such asymmetries. The left hand is not an *exact* mirror-image of the right hand, nor is the left side of the face an exact mirror image of the right side. Even Lewis Carroll, who was fascinated by mirror images, was somewhat asymmetrical, as Martin Gardner confirms: "In appearance Carroll was handsome and asymmetric—two facts that may have contributed to his interest in mirror-reflections. One shoulder was higher than the other, his smile was slightly askew, and the level of his blue eyes was not quite the same."[12] Yet it is also clear that we are much more bilaterally symmetrical than chance would dictate. The two sides of the body are very close indeed to being mirror images, so one must conclude that selective pressures operated to ensure a high degree of bilateral symmetry.

Some of these pressures are not difficult to discern. Given the constraints of biological systems, linear movement is best achieved by having paired limbs, be they legs, wings, or flippers—or oars, for that matter. The most efficient form of movement between two points is a straight line, so there is pressure to have symmetrical limbs. There are occasional discrepancies, such as the asymmetrical gallop of the horse or the sideways crawl of the crab. I did not learn to ice-skate until I was well into adulthood and to this day am only able to turn to the

[12] Gardner 1960, 10.

left—one of the penalties for being an unusually lopsided ape, I suppose. This asymmetry is a great nuisance if the hockey puck happens
to be to my right, and my only route toward it is a circuitous one; I
would clearly not have survived selection into the world of professional hockey. Once the principle of paired limbs is established, there
is pressure for the sense organs also to be paired and symmetrically
placed. It's no use having eyes only on one side of the head—Martin
Gardner tells us why: "The slightest loss of symmetry, such as the loss
of a right eye, would have immediate negative value for the survival
of any animal. An enemy could sneak up unobserved on the right!"[13]
One exception is the flatfish, in which the eyes have migrated to the
same side of the head, for otherwise one eye would be condemned to
stare into the ocean bed.

Bilaterally symmetrical limbs and sense organs would have dictated the symmetry of the brain, at least insofar as it has to do with
the processing of sensory input and the organization of reactions to
the environment. Many actions are largely dependent on input from
the spatial environment. In climbing trees, or picking fruit, or catching
insects, an animal often needs to respond efficiently, or react quickly,
to information on one or the other side. Sensory information on one
side of the environment is projected to the opposite side of the brain,
and movements of the limbs are also controlled largely from the opposite side of the brain. If you reach quickly with your left hand to
catch an object, such as a cricket ball, that flies to your left, both the
perception of the ball and the action of the left hand are organized in
the right side of the brain.[14] Conversely, the act of catching a ball on
the right with the right hand is organized in the left side of the brain.
This arrangement is presumably designed for efficiency, since there is
no neural time lost by having to transfer information between the two
sides of the brain.[15] Assuming that cricket balls
and other things are as likely to appear on the left
as on the right, and that you reach for them with
the nearest limb, it makes sense to have the sensorimotor programming accomplished symmetrically and bilaterally.

But it is also clear that the principle of bilateral symmetry is readily abandoned where it is
less adaptive than some asymmetrical arrange-

[13] Gardner 1967, 70.
[14] Baseball won't do as an illustration here, as I'm told that baseball players wear a mitt on just one hand, and so don't use opposite hands to catch balls on either side. Cricketers don't need mitts, just as real men don't eat quiche.
[15] What is not so obvious is why the mapping is contralateral, that is, why the right side of the brain deals with the left side of space and the body, and the left side deals with the right side. For some speculations as to how and why this might be the case, see Kinsbourne 1978.

ment. Internal organs that are *not* involved with sensory or motor functions tend not to be symmetrical and may even show pronounced deviations from symmetry. These include the heart, stomach, liver, and so on; their somewhat asymmetrical shape or placement is presumably largely a matter of efficient packaging. (You would be foolish, or else possessed of an unusually severe compulsive disorder, if you were always to pack a suitcase with the contents arranged in perfect bilateral symmetry.) Departures from symmetry are also found in the brain in functions that are somewhat removed from environmental input or the organization of actions on the environment. For example, speech might be regarded as a largely autonomous process that is little guided by incoming sensory information, except perhaps sensory feedback from the voice itself, and it would be inefficient to have a complex process like this duplicated in the two hemispheres of the brain or constrained to a symmetrical organization across the two sides of the brain.

Much the same principles have governed the body plan of an automobile, although the selective pressures here probably have as much to do with market survival as with the biological survival of drivers and passengers. An automobile is not symmetrical with respect to top and bottom or back and front, for reasons that have obviously to do with gravity and motion. Although it is bilaterally symmetrical on the outside, its internal organs are not symmetrically located, and the leftward placement of the steering wheel even suggests left-hemispheric responsibility for action. In my own vehicle, though, the steering wheel is on the right, a legacy of my colonial heritage.

■ Cerebral and manual asymmetries in nonhuman species ■

In seeking possible antecedents to right-handedness and left-cerebral representation of language, we are concerned with asymmetries that apply to the majority of the population. These are called *population-level* asymmetries, as distinct from the random or fluctuating asymmetries that occur in all organisms. For example, even mice may show quite a strong preference for one or the other paw in reaching for food in a glass tube, but as many are left-pawed as are right-pawed.[16] This random fluctuation may have little to do with the strong right-hand preference evident in all human societies.

[16] *Collins 1970.*

Although the brain is to a very great extent bilaterally symmetrical in all species, population-level brain asymmetries occur in a variety of vertebrate species, including fish, reptiles, amphibians, and mammals. According to one recent review,[17] all species so far tested that show gregarious behavior display such asymmetries, and the evidence is generally consistent with a left-brained specialization for categorizing things, an activity that might be regarded as a precursor to language. They also show a complementary right-brained specialization for attack and agonistic behavior, which resembles the right-brained specialization in humans for emotion and spatial activity.

Although vocalization appears to be under largely subcortical control in nonhuman species, there is evidence for a left-sided bias, even in the frog.[18] Frogs are anurans, probably the first vertebrates with vocal chords, so a left-hemispheric bias for vocalization may go back to the very origins of vocal behavior some 170 million years ago. Many other examples of left-hemispheric specialization for the production or interpretation of vocal sounds could be cited. In passerine birds, such as chaffinches and canaries, singing is accomplished by an organ called the syrinx, which is analogous to the larynx in humans. If the nerve that stimulates the syrinx on the left is cut, the bird loses most of its song pattern, whereas cutting the nerve on the right has relatively little effect.[19] Since the brain structures controlling this nerve are on the same side, this has been taken to imply left-brained control.[20] The *perception* of vocalizations that are specific to the animal's own species is left-hemispheric in mice, rats, marmosets, rhesus monkeys, and Japanese macaques.[21] This striking evidence for a left-hemispheric specialization for vocalization across a wide range of species, including at least some primates, suggests that this may indeed be a near-universal bias. Of course vocalization is not *language*, but the bias may possibly explain why language became lateralized to the left hemisphere, as I explain below.

Little is known of functional asymmetry in the chimpanzee brain, but some studies have focused on anatomical asymmetries, especially in an area of the brain known as the *temporal planum*, which in humans is part of Wernicke's area and plays a critical role in the comprehension of speech. This area is larger on the left than on the

[17] See Vallortigara, Rogers, and Bisazza 1999.
[18] Bauer 1993. We might take this as welcome evidence that we are not, after all, descended from birds.
[19] Nottebohm 1977.
[20] It has also been argued that the mechanism by which this asymmetry is accomplished may not be comparable to that in humans (Goller and Suthers 1995).
[21] For mice, see Ehret 1987; for rats, see Fitch, Brown, O'Connor, and Tallal 1993; for marmosets, see Rogers 2000; for rhesus monkeys, see Hauser and Anderson 1994; and for Japanese macaques, see Heffner and Heffner 1984.

right in about 65 percent of the population.[22] This has generally been taken to reflect the dominance of the left hemisphere for speech, although some have questioned this interpretation.[23] As we saw in the previous chapter, we have some evidence that Wernicke's area itself was enlarged in the hominins of around 2 million years ago, but not in the australopithecines, although we have no reason to think that this enlargement was greater on one or other side. It therefore came as a surprise when a recent report showed the temporal planum to be larger on the left than on the right in seventeen out of eighteen chimpanzees examined post mortem[24]—a proportion even higher than that observed in humans—although there is some conflicting evidence.[25] The discoverers of this asymmetry point out that it need have nothing to do with vocal communication in the chimpanzee and even suggest that it may relate to a "gestural-visual" mode of communication. Yet, as we shall see below, only about two-thirds of chimpanzees show a preference for the right hand, even in gesturing, and the preference may be restricted to captive chimpanzees. My guess is that the asymmetry may relate to the perception of vocalizations. Recall that Kanzi, the bonobo, is quite skilled at understanding human speech, although probably not to the point of understanding grammar. Nevertheless, to understand words spoken by humans would require sophisticated perceptual processing, and it would not be surprising if this were dependent on specialized mechanisms in the left temporal lobe.

As for handedness, one of the closest parallels to human right-handedness appears to come from the parrot. In most species of parrot, 90 percent of the individuals have a preference for the left foot in picking things up, although one or two species are right-footed.[26] Oddly, vocalization in this talkative bird does not seem to be controlled by the left side of the brain, as it is in passerine birds.[27] The evidence for handedness in primates is mixed. Some species of monkey seem to show a slight predominance of left-handedness in reaching, but not all commentators are convinced by the data.[28] Handedness in monkeys may vary with the task. In one study of capuchins, for example, the ani-

[22] Geschwind and Levitsky 1968.
[23] Jäncke and Steinmetz 1993.
[24] Gannon, Holloway, Broadfield, and Braun 1998.
[25] This was contradicted somewhat in a yet more recent study in which columns of cells in the temporal planum were shown to be more widely spaced on the left than on the right in humans, but not in chimpanzees, an asymmetry weakly reversed in rhesus monkeys (Buxhoeveden and Casanova 2000). An earlier anatomical study by Yeni-Komshian and Benson 1976 showed the Sylvian fissure, which borders the temporal planum, to be longer on the left than on the right in humans. The asymmetry was also present, but less marked, in chimpanzees; it was absent in rhesus monkeys. Wada, Clarke, and Hamm 1975 also found no asymmetry of the temporal planum in rhesus monkeys or baboons.
[26] Rogers 1980.
[27] Nottebohm 1977.
[28] MacNeilage, Studdert-Kennedy, and Lindblom 1987—and see the commentaries following this article.

mals tended to prefer the left hand when reaching into a hole in a box for food, but the right hand when reaching for food items on the ground while supporting themselves with the other hand. When reaching from an upright stance, the animals showed only a slight right-hand preference.[29] Whatever these results mean, they clearly do not reveal the strong right-handedness that most humans would almost certainly show on all three tasks.

As for chimpanzees, William Hopkins found that in a large captive colony about two-thirds of the animals consistently prefer the right hand for a number of activities, such as extracting peanut butter from a tube and, interestingly, gestural communication.[30] But there is no evidence for population-level handedness among chimpanzees or any other primate species in the wild, and the authors of one extensive review concluded that, of all the primates studied, "only chimpanzees show signs of a population bias . . . to the right, but only in captivity and only incompletely."[31]

In sum, the cerebral asymmetry for vocalization may have quite ancient evolutionary roots; it is present in the frog, which is not even blessed with a cerebral cortex! Consistent handedness, in contrast, is present only weakly, if at all, in primates—except, of course, for us dexterous humans. This suggests that cerebral lateralization for language may have originated in the asymmetry for vocalization. That is, once our forebears began to augment gestures with vocalizations, an asymmetry was introduced into the system, creating a leftward bias in the control of communicative gestures as well.

This might have happened roughly as follows. Suppose that language was at one time primarily gestural, but increasingly accompanied by vocalization. These vocalizations would tend more and move to be synchronized with gestures, and in the case of oral gestures, vocalization would create distinctive sounds determined by the gesture itself. The situation might be likened to that of a pianist, whose hand movements bring about sounds by striking the keys. But the piano keyboard is asymmetrical: high notes dominate low notes, at least in the ear of the listener. Since the high notes are on the right, the right hand would be favored, giving rise to the left-hemispheric dominance in hand and finger movements.[32] In

---

[29] Parr, Hopkins, and de Waal 1997.
[30] Hopkins 1996; Hopkins and Leavens 1998.
[31] McGrew and Marchant 1997.
[32] As for the piano, I suspect that the causality actually runs the other way. The keys on the right of the keyboard carry the melody because they are in the higher registers, but this arrangement is a consequence of the greater dexterity of the right hand. I doubt that we became right-handed because of the way the piano is organized.

similar fashion, when vocalization was added as a means of augmenting the gestural repertoire, the asymmetry in the control of the vocal chords might have favored the left hemisphere and the right hand.

Recall that the "mirror neurons" in the monkey, which respond both to grasping actions made by the monkey itself and to perceived actions made by others, are represented in the area corresponding to Broca's area on *both* sides of the animal's brain. In humans, the representations of both vocal language and signed language, as well as the equivalent of mirror neurons, appear to be restricted to the left side of the brain, at least in the great majority of us. Somewhere in the progression from ape to human, vocalization and gesture were linked, and the system became lateralized. My guess is that this happened when vocalization was added to the gestural repertoire and synchronized with it.

Prior to this linkage, it is understandable that vocal control might be more strongly lateralized than manual control. Much of the manual activity of primates has to do with climbing trees and with reaching for things and manipulating them. Access to branches of trees or objects to pluck are as likely to occur on one side of the body as on the other, and it is this very lack of systematic bias that no doubt preserved the symmetry of the limbs and the controlling brain structures in the first place. But vocalization does not depend on the spatial layout of the environment, and little is lost, and perhaps much gained, by having it under asymmetrical control.

But as the hands became involved in communication, they too might have gained from asymmetrical control. Communication and manipulation might be considered operations upon the environment rather than reactions to it, and so the spatial layout of the physical world is less of a constraining influence. As both activities became more complex, symmetry would have become a handicap, creating potential conflict and inefficiency. Bipedalism would, of course, have greatly increased the opportunity for both manipulation (as in the use and manufacture of tools) and complex communication—itself a form of manipulation. In activities of this sort, which have been termed *praxic* activities,[33] it is perhaps not surprising that cerebral control became lateralized.

I have suggested that the left-hemispheric dominance for both manual and vocal control have their origins in an ancient leftward

---

[33] *Corballis 1991.*

asymmetry in vocal control, going back perhaps 170 million years to our common ancestry with birds and amphibians. Yet in humans the two asymmetries are not perfectly concordant, even though around 90 percent of the population are left-brained for both manual and vocal function. There are a few left-handed people who are left-brained for language, and a few right-handed people who are right-brained. This variation is also of interest and may have a genetic basis.

■ Genetic theories of human handedness ■

Handedness does tend to run in families. In one survey, data accumulated from over 70,000 people showed the percentage of left-handers among those born to right-handed parents to be 9.5 percent. This rose to 19.5 percent among those who had one right-handed parent and one left-handed parent, and to 26.1 percent among those born to two left-handed parents.[34] In the words of animal breeders, handedness does not *breed true*; having left-handed parents merely decreases the probability that you will be right-handed, but you are still more likely to be right- than left-handed. The inheritance of lopsidedness is lopsided.

One might be tempted to conclude from this that handedness is simply due to environmental influences. Right-handedness is deeply embedded in our environment and culture, in the way we eat, greet each other, design implements (such as scissors or golf clubs), and place door handles. Even books and magazines are designed for the convenience of the right-handed and the frustration of the lefty. The idea that we are essentially taught to be right-handed does have its advocates,[35] but there are some compelling reasons to believe that the fundamental basis for the bias toward right-handedness is biological rather than social and is a cause rather than a consequence of our right-handed world.

One reason is that handedness is clearly linked to left-cerebral dominance for speech, even though not perfectly concordant with it, and it's hard to see how the lateralization of speech could be environmentally induced. The relation between handedness and brainedness for speech is roughly linear, such that the less right-handed you

[34] *McManus and Bryden 1992.*
[35] *Provins 1997.*

are, the more likely it is that your right brain will be dominant for language. But there is an overall bias in favor of a loquacious left brain: one recent study shows that the percentage of left-brained dominance ranges from 96 percent in extreme right-handers to 73 percent in extreme left-handers.[36] It may come as a surprise that the majority of *left*-handers are also left-brained for language,[37] although the proportion of people with language representation on both sides of the brain appears to be higher among left-handers than among right-handers. Left-handers do not simply have their brains in backward. The relation between handedness and the cerebral representation of signed language is probably similar, although the data are sparse. Doreen Kimura has surveyed cases of manual aphasia—cases in which deaf signers have shown disorders in signing following brain injury—and found that in nine right-handers the damage was to the left side of the brain, but that in two left-handers the damage was to the left side in one case and the right side in the other.[38] These results are at least consistent with the idea that language dominance in left-handers could go either way but is nearly always left-hemispheric in right-handers.

Right-handedness for reaching appears to be demonstrable in infants at the tender age of twenty weeks,[39] and it is unlikely—although perhaps not impossible—that this is induced by parental influences. The leftward asymmetry of the temporal planum is evident as early as the twenty-ninth gestational week,[40] at which age the fetus is as immune to parental influence as is the teenager. Yet a further reason to suppose that right-handedness is fundamentally biological is that it is a human universal. Right-handedness is common to all human cultures, including the indigenous Australians, who were isolated for tens

[36] *Knecht, Dräger, Deppe, Bobe, Lohman, Flöel, Ringelstein, and Henningen 2000 studied dominance for language in healthy individuals, using a technique called functional transcranial Doppler sonography, which records relative activity in the two sides of the brain while people generate words. To measure handedness, they used the Edinburgh Handedness Inventory, which assigns a "laterality index" ranging from −100 (extreme left-handedness) to +100 (extreme right-handedness) based on their hand preferences for such tasks as writing, eating, using a knife, and so on (Oldfield 1971). If you know your laterality index, you can roughly estimate how likely it is that you are right-brained for language by dividing the laterality index by 10, and subtracting it from 15, to give a percentage score. For example, if you score 60, this gives a value of 9 percent, which is an indication of the chances that it's your right brain that does the talking. A score of −60 would raise the chances to 21 percent.*

[37] *This is confirmed by a number of other studies, using a variety of different techniques (Pujol, Deus, Losilla, and Capdevila 1999; Rasmussen and Milner 1977; Warrington and Pratt 1973). The figures may give you the illusion that the bias toward left cerebral dominance is stronger than that toward right-handedness, but this is not necessarily so. Indeed, the relations are likely to hold in reverse: a majority of those with right-cerebral dominance for language are right-handed, although the proportion of right-handers is even higher among those with left-cerebral dominance. Go figure.*

[38] *Kimura 1981.*

[39] *Morange-Majoux, Peze, and Bloch 2000.*

[40] *Wada et al. 1975.*

[41] *Dawson 1972 gives a figure of 10.5 percent for left-handedness among indigenous Australians. A useful table of the incidence of left-handedness across a wide range of ethnic and cultural groups is provided in Porac, Rees, and Buller 1990. While there are variations, the largest percentage for adults is 15.5 percent, recorded in one survey in England, and the lowest is 1.5 percent, recorded in Hong Kong. These variations depend partly on the criteria used to define left- or right-handedness, and partly on cultural norms: in some cultures there are strong sanctions against the use of the left hand for certain activities, such as writing and eating, and sanctions against using the right hand for another activity that shall be nameless.*

[42] *Annett 1972. There are other biological precedents for supposing that the genes influence the presence or absence of asymmetry rather than the direction of an asymmetry. For example, in a mutant strain of mice the asymmetry of the heart is reversed (situs inversus) in precisely 50 percent of the population, and normal in the remaining 50 percent (Brueckner, D'Eustachio, and Horwich 1989). These mice are homozygous for the so-called iv allele, which is therefore analogous to the C allele in the handedness model. Situs inversus in humans, a very rare condition, appears to have a similar genetic basis, although there is no evidence that it has any direct bearing on handedness or cerebral asymmetry.*

[43] *Annett 1993 actually uses the labels "RS+" (right shift present) and "RS−" (right shift absent) to refer to these alleles. I have used instead the terminology proposed by McManus 1985, because it is simpler and allows me to introduce Donald Duck, who will appear shortly. Birds are important to me. McManus's model actually differs in a number of ways from Annett's but shares with it the notion that the alleles signal, not leftward versus rightward biases, but rather the presence versus the absence of a rightward bias.*

[44] *Crow 1998.*

[45] *Corballis 1997.*

[46] *See McKeever 2000 and the exchanges between* (continued)

of thousands of years.[41] It seems unlikely that a cultural pressure toward right-handedness would have persisted essentially unaltered across the extremes of time and cultural variation. Most theorists have indeed accepted that human handedness is biological rather than cultural, and a number have tried to develop genetic theories to explain the variation in handedness between individuals.

The most compelling genetic theories are based on the insightful suggestion of Marian Annett that human handedness may depend on two genetically based influences, one creating a bias toward right-handedness and the other creating no disposition toward either right- or left-handedness.[42] From a genetic point of view, the dichotomy is not between left- and right-handedness but between being right handed *or not*. Imagine, then, a single gene, which Annett calls a "right shift" gene, with two alternative forms, or *alleles*. One allele might be called D, for "dextral," because it codes for a shift of the handedness distribution toward right-handedness, and the other might be called C, for "chance," because it leaves the direction of handedness to random influences.[43]

I should point out that this gene is purely hypothetical, and no such gene has yet been located on the human genome, although there has been some speculation as to where it might be. Timothy Crow has argued that the gene is located in homologous (matching) regions of sex chromosomes.[44] If true, this would at least narrow the search for it, but I have myself argued that it is unlikely to lie on both the X and Y chromosomes,[45] although a case can be made for supposing that it might lie on the X chromosome alone.[46] I need not bore the reader with the details of these arguments, which somewhat resemble theological arguments as to how many angels can dance on the head of a pin. You should understand, though, that in the remainder of this chapter the handedness gene should be understood as hypothetical, with much the same status as the hypothetical "particles" proposed by Mendel to explain how crossbreeding affects the physical characteristics of peas. Only later did these "particles" come to be called "genes" and later still achieve some degree of physical reality. But remember, too, that Mendel turned out to be right.

Let us suppose, then, that there is indeed a gene with two alleles, D and C, that underlie variations in handedness. At conception, we each receive two copies of the gene, one from each parent. Those who receive two copies of the D allele, known as DD homozygotes, will be strongly impelled toward right-handedness. In those who receive one C allele and one D allele, known as CD heterozygotes, there will be a weaker push to the right. In CC homozygotes, who receive two copies of the C allele, handedness is simply a matter of chance. Chris McManus has suggested that the proportions of right-handers in each genotype should be 100 percent for DD homozygotes, 75 percent for CD heterozygotes, and 50 percent for CC homozygotes, and these figures, which are again hypothetical, actually produce a good fit to data on the inheritance of handedness.[47]

It might be supposed, then, that the D allele emerged at some point in hominin evolution and governed not only handedness but also the dominance of the left side of the brain for language and manual control. In effect, it guaranteed that, in the great majority of individuals, handedness and speech control are represented in the same side of the brain. This happy union may well have been selected during the incorporation of vocal elements into gestural language. Doreen Kimura found that right-handers tend to gesture with the right hand while they speak, while left-handers are more variable, showing a greater tendency to gesture with *both* hands.[48] All this is consistent with the view that right-handers are more likely to inherit the D allele, and thus to show consistent lateralization of hand and voice.

My account so far may have emphasized the

Jones and Martin 2000 and Corballis 2001.
[47] McManus 1985. Elsewhere, though, I have argued that the "default" condition in the absence of the D allele may not be equal probabilities of left- and right-handedness, but rather a 67 percent bias in favor of right-handedness (Corballis 1997). This helps explain a number of other weaker asymmetries, such as that of the temporal planum, and is consistent with evidence that great apes show a 67 percent bias in favor of the right hand on some activities. This refinement need not concern us too much here. The important point is that the dextral allele is assumed to be unique to humans, and responsible for the 90 percent right-handedness of our own lopsided species.
[48] Kimura 1973a, 1973b.

dominance of the left side of the brain at the expense of complementary specializations in the two sides. The work on the split brain, alluded to earlier, at least had the beneficial effect of emphasizing that the right hemisphere is in some respects superior to the left, especially in more passive functions such as spatial perception and perception of emotion. But it is still the left hemisphere that provides us with the most dramatic examples of specialization, notably in the left-hemispheric control of speech and the pervasive dominance of the right hand. It is actually quite difficult to demonstrate right-hemispheric advantages in the split brain, and Michael Gazzaniga, a pioneer of human split-brain research, has provocatively suggested that the disconnected right hemisphere might be "vastly inferior to the cognitive skills of a chimpanzee."[49] He has since modified his view somewhat, noting evidence that the right hemisphere is better, in subtle ways, at some perceptual functions,[50] although his claim is not so much that the right hemisphere is better but that the left hemisphere is worse! That is, the left hemisphere has forfeited some of its capacity for perceptual functioning because a lot of its neural circuitry is taken up with the invading presence of language and the "interpreter."[51]

Gazzaniga also suggests that the right hemisphere keeps a veridical record of past events, while the left hemisphere tends to interpret and therefore distort the past. But again, he cannot resist noting that although this sometimes leads to better performance by the right hemisphere, its performance in at least one respect is comparable to that of rats and goldfish.[52] It goes like this. Suppose you get rewarded for guessing which of two possible events will occur. The events happen randomly in sequence, but one is, overall, more likely than the other. You can maximize your reward by always guessing the more likely event, or you can try to match the actual frequencies. The left hemisphere tends to match, but the right hemisphere tends to pick the more likely event, and so do goldfish. Next time you go to the casino, close down your left brain—or else bring a goldfish along for advice.

This aside, not all researchers of the split brain have Gazzaniga's rather poor opinion of right-hemispheric functioning,[53] and the issue is complicated by the fact that it is difficult to fully test the capacities of the disconnected right hemisphere because of

---

[49] Gazzaniga 1987, 535.
[50] See, for example, Funnell, Corballis, and Gazzaniga 1999.
[51] Gazzaniga 2000.
[52] Ibid., 1316.
[53] See Bogen 1993 or Zaidel 1983.

its relatively poor understanding of instructions, which are almost inevitably verbal. Perhaps we simply don't yet know much of what the right hemisphere can do, because we do not have appropriate ways to ask it questions, or even know the right questions to ask. But I think it fair to conclude that those functions that are *distinctively* human, including speech, manufacture, and the complex planning and execution of sequences, are fundamentally dependent on the left hemisphere in most people. My own view, like Gazzaniga's, is that much of the *right* hemisphere's specialization arises because the left hemisphere, with its role in language and sequencing, has simply relinquished some of its more routine duties.[54]

A possible example of this comes from a phenomenon known as hemineglect. People suffering damage to the right side of the brain often suffer from a neglect of the left side of the world. They may eat only from the right side of the plate, talk to people only if they stand to the right, even dress only the right side of the body. If playing chess against a person with hemineglect, you would be advised to attack down the right side of the board, which is to say your opponent's left side, and your attack is likely to go unnoticed. The curious thing is that neglect of the right side of space is hardly ever observed following left-sided brain damage, and when it does occur, it is usually transient.

Hemineglect is traditionally associated with damage to the right parietal lobe of the brain. In monkeys, though, it is damage to the upper portion of the temporal lobe that produces neglectlike phenomena, and damage to the left side produces as much neglect as damage to the right side.[55] But now comes evidence that it is really damage to the right superior temporal lobe in humans that is critical to hemineglect.[56] Many patients with neglect have also suffered damage to the neighboring parietal areas, and this may have misled investigators into thinking that the parietal lobe is important. But if the right superior temporal lobe is actually important for spatial awareness, what is the corresponding area on the left doing? You've guessed it: it is one of the main areas involved in language and is part of Wernicke's area. The invasive presence of language has deprived it of its former role in spatial awareness.[57]

[54] *Corballis and Morgan 1978; and see also Corballis 1998. Gazzaniga 2000 seems to agree.*
[55] *Watson, Valenstein, Day, and Heilman 1994.*
[56] *Karnath, Ferber, and Himmelbach 2001.*
[57] *I can't resist saying I told you so. I made this suggestion in my 1991 book* The Lopsided Ape *(263). And you wouldn't listen.*

■ Why do left-handers survive? ■

But one must then ask why the D allele did not simply replace the C allele altogether, as normally happens in evolution when one allele is associated with greater fitness? To put it more crudely, why do left-handers survive? In 1991 two psychologists, Stanley Coren and Diane Halpern, argued that left-handers do in fact have "decreased survival fitness."[58] They based this in part on the fact that there are proportionately fewer left-handers in older age groups than in younger ones, and in part on other evidence, including an analysis of data from a baseball record book that noted the handedness of players, along with their dates of birth and death. Their conclusion attracted considerable interest and controversy in the media. This is not the place to enter the debate, but readers might like to read a detailed refutation by Lauren J. Harris, a long-time commentator and researcher on handedness.[59] Harris is himself a left-hander, and in correspondence with me over this issue was happy to inform me that he is still alive.[60] I sincerely hope that many octogenarian left-handers are reading these very words.

In any event, it is unlikely that any impediment to survival was associated with the C allele over evolutionary time, since it takes only the slightest difference in fitness between two alleles for the fitter to supersede the less fit. In fact, left-handers appear to have been a stable presence, at roughly 12 percent of the population, for as far back as the historical record takes us.[61] The most likely reason for the stability of variation in handedness is that C and D alleles have been sustained in the population by a so-called *heterozygotic advantage*. That is, CD heterozygotes are slightly fitter than either CC or DD homozygotes, and this is sufficient to ensure that both alleles remain in a "balanced polymorphism."

The idea of a heterozygotic advantage is well established among animal breeders as a mechanism for increasing resistance to disease and overall fitness; they call it *hybrid vigor*. A well-known example has to do with a hemoglobin gene, with one allele that causes sickle-cell anemia. The sickle-cell allele is recessive, so that only those who carry two copies of it develop the disease. Those heterozy-

[58] *Coren and Halpern 1991.*
[59] *Harris 1993.*
[60] *Stanley Coren is a right-hander, and has received threats to his own survival for daring to suggest that left-handers die earlier than right-handers. As an anonymous and presumably sinister caller put it to him, the statistics can be altered. I should also add that Lauren Harris is not, to my knowledge, an octogenerian.*
[61] *See, for example, Coren and Porac's survey of handedness (1977) as depicted in art over the past five thousand years.*

gotes who carry one sickle-cell allele and one allele for normal hemoglobin do not develop the disease and are more resistant to malaria than are those homozygotic for normal hemoglobin. This heterozygotic advantage has preserved the sickle-cell allele in African populations where malaria is endemic, despite the fact that those unlucky enough to be homozygotic for the allele inevitably die of anemia.

I do not mean to imply that the C allele is at all comparable to the sickle-cell allele, despite Coren's suggestion that left-handers have lower survival chances. Nevertheless, it may well be the case that there are advantages to being equipped with both C and D alleles. To help the reader keep the genotypes in mind, then, let me introduce the DD genotype as Donald Duck, and the CC genotype as Charlie Chaplin— a left-hander, as one might expect. Who could the CD be but the evolutionarily fittest of them all? Charles Darwin, of course. We now need to explore why CD might be slightly fitter than DD or CC.

◼ On the fitness of genotypes ◼

One view, proposed by Marian Annett, is that DD genotypes are likely to be verbally fluent but deficient in spatial skills—all quack and no direction. The reason for this, she suggests, is that left-cerebral dominance is achieved by a pruning of the right hemisphere during development, and it is the right hemisphere that in most people is the more proficient in spatial orientation and other nonverbal skills. DD may be a great orator, but tends to get lost on the way to the forum. The absence of any pruning mechanism in CC individuals may lead to superior spatial skills but creates the risk of verbal dysfunction—as befits a Chaplinesque role in silent movies, perhaps. The ideal mix then might be provided by the CD genotype, ensuring a balance of verbal and spatial skills.

According to this theory, left-handers are of course more likely than right-handers to be CC genotypes, and there have long been suggestions that left-handedness or the lack of consistent dominance are associated with language-related disorders such as reading disability[62] and stuttering.[63]

[62] This theory is often associated with the work of Samuel Torrey Orton in the 1920s and 1930s, and summarized in his book Reading, Writing, and Speech Problems in Children 1937.

[63] Orton was also associated with the view that stuttering is due to lack of cerebral dominance, but it was his colleague Lee Edward Travis who elaborated it, especially in his book Speech Pathology 1931. But Travis later changed his mind, and his successor at the University of Iowa, Wendell Johnson, specifically repudiated the so-called "dominance theory" (Johnson 1955). Yet there is still occasional evidence for it. For example, several stutterers have reportedly been cured of their affliction following unilateral brain surgery, which suggests that the stuttering was caused by conflict between the two sides of the brain (see, e.g., Andrews, Quinn, and Sorby 1972).

The evidence has been decidedly mixed on these claims, and perhaps more negative than positive, although it should be remembered that the putative CC genotype merely increases the risk of inconsistent dominance, and perhaps only slightly; environmental biases will normally establish asymmetry, as in the random determination of handedness in mice.

Perhaps the most impressive evidence on the relation of handedness to reading and other academic skills comes from an examination of test scores for 12,770 individuals in a British national cohort.[64] Their handedness was measured on a continuous scale from extreme left to extreme right. Their scores on tests of verbal ability, nonverbal ability, reading comprehension, and mathematical ability all showed a pronounced dip right on the point of equality between the hands. That is, left- and right-handers were largely indistinguishable, but those with equal hand skill had lower scores. The authors called this "the point of hemispheric indecision." This gives support to the idea that CC individuals are at risk for deficits in academic abilities—and not just verbal ones. This risk is quite small, since chance influences will cause most CC individuals to show consistent handedness, and the effect is likely to be missed in studies that simply compare left- and right-handers.[65]

The risk, of course, would be eliminated in DD individuals, higher in CD individuals, but highest in CC individuals. But this raises once again the question of why the C allele has persisted, especially since the dip at the point of hemispheric indecision was not confined to language-related abilities, but included a nonverbal test that involved the processing of shapes. Here's where we need a little magic, perhaps of a Chaplinesque variety.

[64] *Crow, Crow, Done, and Leask 1998.*

[65] *This account is actually more consistent with Annett's treatment than with McManus's. McManus regards handedness as fundamentally dichotomous, so that CC individuals are no more likely to be ambilateral than CD or DD individuals. Annett, by contrast, regards handedness as lying on a continuum from extreme left to extreme right. In this model, those lacking the "right-shift" allele, here referred to as CC individuals, are more likely to be ambilateral than those who have one or two copies of this allele (i.e., CD and DD individuals). With apologies to Marian Annett, I continue to use the labels C and D in order to appease the insistent Duck.*

■ Magic on the brain ■

A number of researchers have suggested that the lack of cerebral asymmetry somehow encourages what has been called "magical ideation." This refers to beliefs in phenomena like extrasensory perception, telekinesis, extraterrestrial invaders, and other phenomena that defy the normal laws of physical causation or common sense. Such beliefs are common in mental disturbances such as schizophrenia, where sufferers may be convinced that some agency is trying to implant ideas into their minds, whether by extrasensory means or through some surreptitious technology.

Mixed handedness for skilled tasks has been reported as higher among those assessed as having schizotypal personality traits,[66] and there is also some evidence that schizophrenics themselves show an elevated incidence of mixed or ambiguous handedness.[67] Timothy Crow goes so far as to suggest that "schizophrenia is the price that *Homo sapiens* pays for language."[68] Magical ideation seems to be related to hemispheric asymmetry, as measured by comparing people's ability to make verbal decisions about words flashed to the left or right of where they are looking. Because of the rather peculiar way in which the brain is wired up, visual objects that appear to the left of where you are looking are projected to the right side of the brain, while objects on the right are projected to the left side of the brain. People are therefore usually better at processing words if they are flashed to the right of where they are looking, since the words are then projected to the left side of the brain, which is the dominant side for language. Words flashed to the left side are projected to the right side and are disadvantaged, because the right brain has only limited verbal capacity. Experiments have shown that people with a strong belief in extrasensory perception[69] and those who scored high in magical ideation[70] did not show the expected right-sided (left-brained) advantage when asked to decide whether strings of letters to the left or right of where they were looking.

Figure 8.1 shows scores on a magical ideation questionnaire plotted against relative hand preference.[71] What is striking is that magical ideation peaks precisely at the "point of hemispheric indecision." This suggests that the D and C alleles balance not so much our verbal and spatial abilities as our capacities for rational and magical thinking. The C allele has remained in balance with the D allele because it injects a little magic into our lives.

It may seem odd that magical thinking should have an influence on reproductive fitness, which is what evolutionary selection is all about. One possibility is that it has to do with sexual selection rather than natural selection. Magical ideation may

[66] Poreh, Levin, Teves, and States 1997.
[67] Shan-Ming, Flor-Henry, Dayi, Li Tiang, Qui Shuguang, and Ma Xenxiang 1985; Shimizu, Endo, Yamaguchi, Torii, and Isaka 1985.
[68] Crow 1997.
[69] Brugger, Gamma, Muri, Schaffer, and Taylor 1993.
[70] Leonhard and Brugger 1998.
[71] Barnett and Corballis 2002.

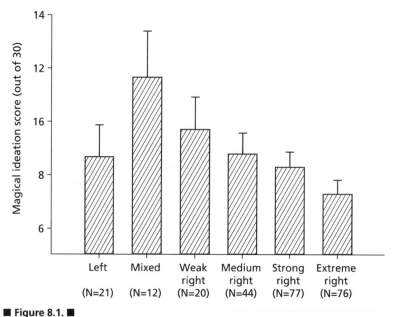

■ **Figure 8.1.** ■

Scores on a Magical Ideation Scale, plotted against degrees of handedness (see Barnett and Corballis 2002). The vertical lines represent standard errors of the mean.

simply be sexy, as is the peacock's tail to the peahen, although the burning of witches suggests selection against magical powers, rather than selection for them. Even so, there can be no question that magical ideation is a prominent part of our lives, as we scrutinize horoscopes, carry lucky charms, avoid walking on the cracks in the pavement, or read Harry Potter books. Magical thinking is a feature of most religions, which typically include belief in supernatural powers, life after death, faith healing, and the ability to influence the future through prayer. Religion remains very much a part of everyday life in probably the majority of the world's population, although its influence has declined in industrialized countries. Yet even in the United States, only around 3 percent of the public describe themselves as agnostic or atheist.[72] Some research indicates that religion and spirituality are beneficial to both mental and physical health, and therefore perhaps to survival,[73] although the evidence is not all positive. Some religions are overly authoritarian or literal and have been associated with child abuse and neglect,

[72] *Thoreson 1999. One group that differs markedly from the public norm is that of health professionals, some 50 percent of whom describe themselves as agnostic or atheist. Perhaps this simply represents a tendency to take upon themselves Godlike properties.*

[73] *Larson, Swyers, and McCullough 1998.*

and—harking back to Jaynes—with false perceptions of control.[74] Religion is often seen as in conflict with rational or scientific thinking, not least with respect to the theory of evolution itself, but perhaps a case could be made for the idea that the two are kept in balance by complementary survival demands. It's an odd thought that the theory of evolution itself may lack survival value.

Of course, magical thinking is often associated with schizophrenia, which certainly creates doubt as to *its* survival value. Yet it may also be associated with creativity, itself often associated with insanity. In 1871 the famous Victorian psychiatrist Henry Maudsley wrote: "I have long had the suspicion that mankind is indebted for much of its individuality and for certain forms of genius to individuals with some predisposition to insanity. They have often taken up the by-paths of thought, which have been overlooked by more stable intellects." The nineteenth-century Italian anthropologist Cesare Lombroso went even further, insisting that genius and madness were essentially two sides of the same coin. In an attempt to prove his theory, he visited Leo Tolstoy, whom he considered the greatest novelist of the day, expecting to find a half-crazed, degenerate individual. But Tolstoy was far from degenerate-looking, and the two men quarreled over another of Lombroso's theories, that criminals are born, not made. Lombroso took Tolstoy's violent objections to this theory as proof of his instability, but one must wonder which of the two was in fact the crazier.[75]

Nevertheless, it is true that many creative individuals have struggled with mental illness, such as the playwright August Strindberg, the New Zealand novelist Janet Frame, the painter Vincent van Gogh, or the composer Maurice Ravel. Where should we locate Albert Einstein on the continuum from magical to rational thinking? It is sometimes said that Einstein was left-handed, but photographs suggest that he was in fact right-handed. Nevertheless, he was said to have been late in developing speech and to have been a slow learner as a child.[76] His thought processes were based largely on visual imagery, and he had difficulty turning his profound thoughts about relativity into mathematical form. While most brains show a left-right asymmetry of the rearward part of the sylvian fissure, which separates the temporal lobe from the central and parietal regions of the brain, Einstein's brain shows "unusual symmetry between the hemispheres" in this region.[77] Einstein may have been perilously close to "the

[74] *Paloutzian and Kirkpatrick 1995.*
[75] *Mazzarello 2001.*
[76] *Highfield and Carter 1993.*
[77] *Witelson, Kigar, and Harvey 1999, 2151.*

point of hemispheric indecision," which may have partly prompted his famous remark, in a 1969 letter to Max Born, that "God does not play dice."

And here's what Einstein had to say about religion: "It is very difficult to elucidate this [cosmic religious] feeling to anyone who is entirely without it. . . . The religious geniuses of all ages have been distinguished by this kind of religious feeling, which knows no dogma. . . . In my view, it is the most important function of art and science to awaken this feeling and keep it alive in those who are receptive to it."[78]

The idea that there might be a dichotomy between rational, logical thinking and intuitive, creative thinking is an old one, now commonly associated with the left and right hemispheres of the brain. I have myself been critical of this dichotomy,[79] which arose from studies of people who had undergone split-brain surgery for the relief of intractable epilepsy. The contrast between left- and right-brain functions has been grossly exaggerated in the popular press and exploited by therapists and educators anxious to release the creative, lateral-thinking right hemisphere from suppression by the left-hemispheric emphasis in our schools and in Western culture generally.[80] Therapists offering right-hemisphere release have not totally forsaken Western values, since they are not averse to charging a healthy fee for their services.

The distinction I am proposing here, though, is not between left- and right-hemisphere processing, but rather between relatively lateralized versus unlateralized brains. I suggest that hemispheric asymmetry itself may lead to more decisive and controlled action, and perhaps a better ability to organize hierarchical processes, as in language, manufacture, and theory of mind. Those individuals who lack cerebral asymmetry may be more susceptible to superstition and magical thinking, but may be more creative and perhaps more spatially aware. The balance between these extremes is preserved by the C and D alleles. The balance is played out not only at the level of individual selection but within society as a whole, in the competing spheres of faith and reason.

With the idea that the lack of consistent asymmetry may be associated with magical thinking,

[78] Quoted in Ramachandran and Blakeslee 1999, 174.
[79] Corballis 1999a—this is a chapter in a book entitled Mind Myths, which deals with a number of popular misconceptions about the brain. Buy it.
[80] A recent example comes from a magazine called Your Horse (no. 211). Apparently you can get along better with your horse if you appreciate that people are left-brained and horses are right-brained. It's nice to get it straight from the horse's mouth.

we have in effect come full circle back to Jaynes, although with not quite the same evolutionary spin. The modern human condition may indeed be a struggle between the bicameral and unicameral minds. It is unlikely, though, that the unicameral mind evolved only as recently as the second millennium B.C.; rather, it developed as a consequence of the greater demands placed on programmed action over the past 2 million years. But could the D allele have emerged 150,000 years ago, in Eve, our founding mother? Timothy Crow has proposed that the appearance of the gene that gives one hemisphere dominance over the other was the "speciation event" that created modern *Homo sapiens*, endowing us with language, cerebral asymmetry, theory of mind—and the risk of psychosis.[81] This bold hypothesis perhaps invests too much in a single mutation; in essence, he argues, all it took for us to graduate from ape to human was the throw of a die some 170,000 years ago. Nevertheless, it is in keeping with the views of those other theorists, such as Philip Lieberman[82] and Derek Bickerton,[83] who have argued that true language appeared late and relatively suddenly, perhaps with the emergence of our own species.

My own view is that language developed much more gradually, starting with the gestures of apes, then gathering momentum as the bipedal hominins evolved. The appearance of the larger-brained genus *Homo* some 2 million years ago may have signaled the emergence and later development of syntax, with vocalizations providing a mounting refrain. What may have distinguished *Homo sapiens* was the final switch from a mixture of gestural and vocal communication to an autonomous vocal language, embellished by gesture but not dependent on it. This switch may well have been facilitated by the emergence of the D allele, which guaranteed that in the majority of people the control of hand and vocalization would be lodged in the same cerebral hemisphere. The Neanderthals, who stayed with us until some 30,000 years ago, may have lacked this final nudge that converted gesture to autonomous speech. Or perhaps, when *H. sapiens* moved into their territory, the Neanderthals simply waited for the Gods to tell them what to do, with disastrous results.

But now the question is, why was the switch to vocal language so important?

[81] Crow 1998. Crow has also suggested that the laterality gene is located in homologous regions of the X and Y chromosomes. I have argued against this on the grounds that polymorphisms are unstable on the Y chromosome, so that either the C or the D allele would eventually disappear from that chromosome (Corballis 1997).
[82] P. Lieberman 1998.
[83] Bickerton 1995.

In the gestures, in the sighs,
Ev'ry day a little dies.
—Stephen Sondheim

9 ■ From Hand to Mouth ■

Imagine trying to teach a child to talk without using your hands or any other means of pointing or gesturing. The task would surely be impossible. There can be little doubt that bodily gestures are involved in the development of language, both in the individual and in the species. Yet, once the system is up and running, it can function entirely on vocalizations, as when two friends chat over the phone and create in each other's minds a world of events far removed from the actual sounds that emerge from their lips. My contention in this book has been that the vocal element emerged relatively late in hominin evolution. If the modern chimpanzee is to be our guide, the common ancestor of 5 or 6 million years ago would have been utterly incapable of a telephone conversation but would have been able to make voluntary movements of the hands and face that could at least serve as a platform upon which to build a language.

In chapter 7 I reviewed evidence that the vocal machinery necessary for autonomous speech developed quite recently in hominin evolution. Grammatical *language* may well have begun to emerge around 2 million years ago but would at first have been primarily gestural, though no doubt punctuated with grunts and other vocal cries that were at first largely involuntary and emotional. The complex adjustments necessary to produce speech as we know it today would have taken some time to evolve, and may not have been complete until some 170,000 years ago, or even later, when *Homo sapiens* emerged to grace, but more often disgrace, the planet. These adjustments may have been incomplete even in our close relatives the Neanderthals; arguably, it was this failure that contributed to their demise.

The question now is, what were the selective pressures that led to the eventual dominance of speech? On the face of it, an acoustic medium seems a poor way to convey information about the world; not for nothing is it said that a picture is worth a thousand words. Moreover, we have seen that signed language has all the lexical and grammatical complexity of spoken language. Primate evolution is itself a testimony to the primacy of the visual world. We share with monkeys a highly sophisticated visual system, giving us three-dimensional information in color about the world around us, and an intricate system for exploring that world through movement and manipulation. Further, in a hunter-gatherer environment, where predators and prey are of major concern, there are surely advantages in silent communication, since sound acts as a general alert. And yet we came to communicate about the world in a medium that in all primates except ourselves is primitive and stereotyped—and noisy.

Before I consider the pressures that may have favored vocalization over gestures, it should be repeated that the switch from hand to mouth was almost certainly not an abrupt one. In fact, manual gestures still feature prominently in language; as we have seen, even fluent speakers gesture almost as much as they vocalize, and of course deaf communities spontaneously develop signed languages. It has also been proposed that speech itself is in many respects better conceived as composed of gestures rather than sequences of those elusive phantoms called phonemes.[1] In this view, language evolved as a system of gestures based on movements of the hands, arms, and face, including movements of the mouth, lips, and tongue. It would not have been a big step to add voicing to the gestural repertoire, at first as mere grunts, but later articulated so that invisible gestures of the oral cavity could be rendered accessible, but to the ear rather than the eye. There may therefore have been a continuity from a language that was almost exclusively manual and facial, though perhaps punctuated by involuntary grunts, to one in which the vocal component has a

[1] *See, for example, Liberman and Whalen 2000.*

■ **Figure 9.1.** ■
A pitcher is worth a thousand words.

much more extensive repertoire and is under voluntary control. The essential feature of modern expressive language is not that it is purely vocal, but rather that the vocal component can function autonomously and provide the grammar as well as the meaning of linguistic communication.

What, then, are the advantages of a language that can operate autonomously through voice and ear, rather than hand and eye?[2]

■ Why speech? ■

Advantages of Arbitrary Symbols

One possible advantage of vocal language is its arbitrariness, as I anticipated in chapter 6. Except in rare cases of onomatopoeia, spoken words cannot be iconic, and they therefore offer scope for creating symbols that distinguish between objects or actions that look alike or might otherwise be confusable. The names of similar animals, such as cats, lions, tigers, cheetahs,

[2] The arguments to follow have also been made by a number of authors, including Armstrong 1999 and Givón 1995—and indeed myself (Corballis 1991, 1992, 1999b).

lynxes, and leopards, are all rather different. We may be confused as to which animal is which, but at least it is clear which one we are talking about. The shortening of words over time also makes communication more efficient, and some of us have been around long enough to see this happen: *television* has become *TV* or *telly*, *microphone* has been reduced to *mike* (or *mic*), *autoimmune deficiency syndrome* has dwindled to *AIDS*, and so on. The fact that more frequent words tend to be shorter than less frequent ones was noted by the American philologist George Kingsley Zipf, who related it to a principle of "least effort."[3] So long as signs are based on iconic resemblance, the signer has little scope for these kinds of calibration.

It may well have been very important for hunter-gatherers to identify and name a great many similar fruits, plants, trees, animals, birds, and so on, and attempts at iconic representation would eventually only confuse. Jared Diamond observes that people living largely traditional lifestyles in New Guinea can name hundreds of birds, animals, and plants, along with detailed information about each of them. These people are illiterate, relying on word of mouth to pass on information, not only about potential foods, but also about how to survive dangers, such as crop failures, droughts, cyclones, and raids from other tribes. Diamond suggests that the main repository of accumulated information is the elderly. He points out that humans are unique among primates in that they can expect to live to a ripe old age, well beyond the age of child bearing (although perhaps it was not always so). A slowing down of senescence may well have been selected in evolution because the knowledge retained by the elderly enhanced the survival of their younger relatives.[4] An elderly, knowledgeable granny may help us all live a little longer, and she can also look after the kids.[5]

In the naming and transmission of such detailed information, iconic representation would almost certainly be inefficient; edible plants or berries could be confused with poisonous ones, and animals that attack confused with those that are benign. This is not to say that gestural signs could not do the trick. As we have seen, manual signs readily become conventionalized and convey abstract information. Nevertheless, there may be

[3] *Zipf 1935. He is also famous for Zipf's law, which states that there is a constant relation between where a word ranks in a frequency list and the frequency itself. That is, if you take any word and multiply its rank by its frequency, the result is approximately constant. I bet you'd never have guessed.*

[4] *J. Diamond 2001.*

[5] *Grandad doesn't seem quite so useful, especially since he gave up rugby, and anyway men tend not to live as long as women do. And he needs Granny to look after him too. As Francis Bacon put it, "Wives are young men's mistresses, companions for middle age, and old men's nurses" (from his 1625 essay, Of Marriage and the Single Life).*

some advantage to using spoken words, since they have virtually no iconic content to begin with, and so provide a ready-made system for abstraction.

I would be on dangerous ground, however, if I were to insist too strongly that speech is *linguistically* superior to signed language. After all, students at Gallaudet University seem pretty unrestricted in what they can learn; signed language apparently functions well right through to university level—and still requires students to learn lots of vocabulary from their suitably elderly professors. It is nevertheless true that many signs remain iconic, or at least partially so, and are therefore somewhat tethered with respect to modifications that might enhance clarity or efficiency of expression. But there may well be a trade-off here. Signed languages may be easier to learn than spoken ones, especially in the initial stages of acquisition, in which the child comes to understand the linking of objects and actions with their linguistic representations. But spoken languages, once acquired, may relay messages more accurately, since spoken words are better calibrated to minimize confusion. Even so, the iconic component is often important, and as I look over the quadrangle outside my office I see how freely the students there are embellishing their conversations with manual gestures. Or maybe they're demonstrating about something.

In the Dark

Another advantage of speech over gesture is obvious: we can use it in the dark! This enables us to communicate at night, which not only extends the time available for meaningful communication but may also have proven decisive in the competition for space and resources. We of the gentle species *Homo sapiens* have a legacy of invasion, having migrated out of Africa into territories inhabited by other hominins who migrated earlier. I suggested in chapter 7 that a late wave of *H. sapiens* from Africa may have replaced, not only *Homo erectus* and the Neanderthals, but also groups of *H. sapiens* who had arrived earlier. That word "replaced" may hide a bloodier reality, as I indicated in chapter 7. Perhaps it was the newfound ability to communicate vocally, without the need for a visual component, that enabled our forebears to plan, and even carry out, invasions at night, and so vanquish the earlier migrants. The poet Matthew Arnold, using vocal language, evokes the scene:

And we are here as on a darkling plain
Swept with confused alarms of struggle and flight,
Where ignorant armies clash by night.[6]

It is not only a question of being able to communicate at night. We can also speak to people when objects intervene and you can't see them, as when you yell to your friend in another room. All this has to do, of course, with the nature of sound itself, which travels equally well in the dark as in the light and wiggles its way around obstacles. A wall between you and the base drummer next door may attenuate the sound but does not completely block it. Vision, on the other hand, depends on light reflected from an external source, such as the sun, and is therefore ineffective when no such source is available. And the light reflected from the surface of an object to your eye travels in rigidly straight lines, which means that it can provide detailed information about shape but is susceptible to occlusion and interference. In terms of the sheer ability to reach those with whom you are trying to communicate, words speak louder than actions.

Listen to Me!

Speech does have one disadvantage, though: it is generally accessible to those around you and is therefore less convenient for sending confidential or secret messages or for planning an attack on enemies within earshot. To some extent, we can overcome this impediment by whispering. And sometimes, people resort to signing; we saw in chapter 6 that signed languages are often used to overcome speech taboos or vows of silence, as among the indigenous Australians of the North Central Desert or religious communities. But the general alerting function of sound also has its advantages. When Mark Antony cried, "Friends, Romans, countrymen, lend me your ears," he was trying to attract attention as well as deliver a message.

In the evolution of speech, the alerting component of language might have consisted at first simply of grunts that accompany gestures to give emphasis to specific actions or encourage reluctant offspring to attend while a parent lays down the law. It is also possible that nonvocal sounds accompanied gestural communication. Russell Gray has suggested to me that clicking one's fingers, as children

[6] *From* Dover Beach *1867. But as a younger chap he had a more romantic view of the night, as in* Faded Leaves *1855:*
Come to me in my dreams, and then
By day I shall be well again!
For then the night will more than pay
The hopeless longing of the day.

often do when putting their hands up in class to answer a question, may be a sort of "missing link" between gestural and vocal language. I know of no evidence that chimpanzees or other nonhuman primates are able to click their fingers as humans can, although lip smacking, as observed in chimpanzees, may have played a similar role. Sounds may therefore have played a subsidiary and largely alerting role in the early evolution of language, gradually assuming more prominence in conveying the message itself.

Of course visual signals *can* capture attention. Indeed peripheral vision is primarily dedicated to the detection of movement or sudden changes in illumination that might signal danger or events of special interest. But peripheral vision extends somewhat less than a quarter circle to the left or right of where one is looking. Other mammals, such as horses, have a much wider arc of vision, extending nearly the full circle around them, and perhaps we'd have been better off if we had retained this feature—or at least been equipped with rear-vision mirrors. The human eyes, like those of other primates, face forward, presumably to provide the visual overlap necessary for binocular stereoscopic vision. Some birds have it both ways; they are endowed with two visual systems, a narrow binocular one for close vision when pecking small objects and a monocular one for panoramic vision when flying around seeking prey.

For us mere humans, then, visual signals can only attract attention if they occur within a fairly restricted region of space, whereas the alerting power of sound is more or less independent of where its source is located relative to the listener. And sound is a better alerting medium in other respects as well. No amount of gesticulation will wake a sleeping person, whereas a loud yell will usually do the trick. The alerting power of sound no doubt explains why animals have evolved vocal signals for sending messages of alarm. Notwithstanding the peacock's tail or the parrot's gaudy plumage, even birds prefer to make noises to attract attention, whether in proclaiming territory or warning of danger. Visual signals are relatively inefficient because they may elude our gaze, and in any case we can shut them out by closing our eyes, as we do automatically when we sleep. Our ears, in contrast, remain open and vulnerable to auditory assault, although again some birds have the answer. Ostriches can close their ears, a trick that is useful in a sandstorm, and if it doesn't succeed in blocking out unwanted sounds, they can always go and bury their heads in the sand.

Speech has another, and subtler, attentional advantage. Manual gesture is much more demanding of attention, since you must keep your eyes fixed on the gesturer in order to extract her meaning, whereas speech can be understood regardless of where you are looking. There are a number of advantages in being able to communicate with people without having to look at them, quite apart from the fact that you might find them unattractive—or they you. You can close your eyes in the company of a tiresome companion and still pick up the gist of the discourse, at least until sleep intervenes. You can pretend to be listening to a bore at a cocktail party, but in fact tune your ears to a more interesting conversation elsewhere. More importantly, perhaps, you can effectively divide attention, using speech to communicate with a companion while visual attention is deployed elsewhere,[7] perhaps to watch a football game or to engage in some joint activity, like building a boat. Indeed, the separation of visual and auditory attention may have been critical in the development of pedagogy, but more on this below.

Three Hands Better than Two?

Another reason why vocal language may have arisen is that it provides an extra medium. We have already seen that most people gesture with their hands, and indeed their faces, while they talk. One might argue, then, that the addition of a vocal channel provides additional texture and richness to the message. The combination of hand and voice makes language a son-et-lumière production, as I remarked earlier.

But it is perhaps not simply a matter of more being better. Susan Goldin-Meadow and David McNeill suggest that speech may have evolved because it allowed the vocal and manual components to serve different and complementary purposes.[8] Speech is perfectly adequate to convey syntax, which has no iconic or mimetic aspect, and can relieve the hands and arms of this chore. The hands and arms are, of course, well adapted to providing the mimetic aspect of language, indicating in analogue fashion the shapes and sizes of things, or the direction of movement, as in the gesture that might accompany the statement "He went that-a-way." By allowing the voice to take over

<hr>

[7] *Again, see Givón 1995.*
[8] *Goldin-Meadow and McNeill 1999.*

the grammatical component, the hands are given free rein, as it were, to provide the mimetic component.

Goldin-Meadow and McNeill were careful not to imply, however, that gestural language is somehow deficient in conveying both the grammatical and mimetic components of language. There is little evidence that this is the case. After all, the signed languages of the deaf readily convey both the grammatical and mimetic aspects of communication with little apparent loss of speed or expressiveness relative to speech. As we saw in chapter 5, manual gestures readily intervene if speech is prevented, and even take over some of the syntactic elements.[9] Gesture clearly lurks below the surface of speech, as though ready to come to the rescue when speech fails. And it's usually present even when not required. People gesture while talking on their cell phones, even though their gesticulations fall on blind eyes.[10] Radio broadcasters also gesture, but their messages suffer little from our inability to see them. Indeed, we are perhaps somewhat better off not seeing their gestures; as the Irish broadcaster Terry Wogan observed, "Television contracts the imagination and radio expands it."[11]

Speech may therefore have evolved, not because it gave the hands freer rein for mimetic expression, but rather because it freed the hands for *other* activities. Charles Darwin, who seems to have thought of almost everything, wrote, "We might have used our fingers as efficient instruments, for a person with practice can report to a deaf man every word of a speech rapidly delivered at a public meeting; but the loss of our hands, while thus employed, would have been a serious inconvenience."[12] It would clearly be difficult to communicate manually while holding an infant, or driving a car, or carrying the shopping, yet we can and do talk while doing these things—although one cannot but be struck by the ingenuity of users of signed language in getting around this problem. But perhaps the more compelling advantage of speech has to do with its role in explaining manual techniques.

As we saw in chapter 3, there is evidence that chimpanzees do have techniques for transmitting information about tool use between generations, and that this has given rise to distinctive cultures

[9] *Goldin-Meadow, McNeill, and Singleton 1996.*

[10] *Actually, though, I suspect that the cell phone may contribute to the ultimate demise of gestures accompanying speech. It is the right hand that plays the leading role in the gestures accompanying speech, at least among right-handers (Kimura 1973a), but most people use the right hand to hold the phone to their right ear, which greatly restricts the opportunity to gesture with that hand. Some people seem to get around this by holding the phone in the left hand, or by propping it against the ear with the shoulder. Women seem to be better at this than men are.*

[11] *Listed in the London Observer of 30 December 1984 as one of the sayings of the year.*

[12] *Darwin 1896, 89.*

in different chimpanzee communities. Sometimes, as when infant chimpanzees in the Taï forest learn how to crack nuts, adults apparently try deliberately to "scaffold" the infants' learning, while in other situations, as when infants in the Gombe National Park learn to fish for termites, the adults appear to offer little explicit help, and the infants learn by simply observing and imitating their elders.[13] Patricia Greenfield and her colleagues note some striking parallels between the ways chimpanzees transmit skills between generations and the way weaving techniques are transmitted by people in Zinacantan, a Mayan community in Ciapas, Mexico. Unlike the chimpanzees, however, Zinacantan mothers use language to help transmit the skills. In doing so, they scaffold their language according to the learner's level of experience and understanding. With young children, for example, they do not try to explain the process but simply tell the child what to do. But with more complex manufacture explanation becomes critical. It would be very difficult indeed to explain how to construct an automobile or a spaceship, or program a computer, without recourse to language and some of its more exotic products, such as mathematics and computer modeling. In that increasingly imperiled institution, the university, instruction is almost entirely verbal, save for a few remaining laboratories.

Speech would have the advantage over manual gesture in that it can be accomplished in parallel with manual demonstrations. Demonstrations might themselves be considered gestures, of course, but the more explanatory aspects of pedagogy, involving grammatical structure and symbolic content, would interfere with manual demonstration if they too were conveyed manually. Clearly, it is much easier and more informative to talk while demonstrating than to try to mix linguistic signs in with the demonstrations. This is illustrated by any good TV cooking show, where the chef is seldom at a loss for either words or ingredients. It may not be far-fetched to suppose that the selective advantages of vocal communication emerged when the hominins began to develop a more advanced tool technology, and they could eventually verbally explain what they were doing while they demonstrated toolmaking techniques. Moreover, if vocal language did not become autonomous until the emergence of *Homo sapiens,* or even later, as I suggested in chapter 7, this might explain why tool manufacture did not really begin to develop true diversity and sophis-

[13] *Greenfield, Maynard, Boehm, and Schmidtling 2000.*

tication, and indeed to rival language itself in these respects, until within the last 100,000 years.

The practical advantages of being able to communicate while using the hands for other purposes may initially have tended to shift the burden of communication from the hand to the face. Facial gestures include the use of the mouth and tongue, and perhaps the addition of extra muscles innervating the tongue, as documented in chapter 7, had to do at first with extending the range of visible gestures of the tongue rather than with speech itself. But gestures associated with the mouth and tongue also have the potential to create sound, independently of voicing, as in the teeth chattering or lip smacking of the chimpanzee, or the click sounds of the Khoisan. And, of course, we can whisper articulately without using the voice at all. But the range of oral gestures could be increased further by making them audible rather than visible. Gestures at the back of the throat or in a closed mouth are invisible but can be rendered audible by adding voicing, or by other clickety-click tricks of the tongue.[14] The changes to the vocal tract that gave rise to speech may therefore have been driven primarily by the advantages associated with freeing the hands from gestural duty through increasing the range of oral gestures.

■ Speech and the rise of technology ■

As we saw in chapter 4, stone tool technology can be traced back to *Homo rudolfensis* some 2.5 million years ago, with the emergence of the Oldowan stone industry. This was followed by the more sophisticated Acheulian industry associated with *Homo ergaster* and African *Homo erectus*, and this, in turn, by the Mousterian industry during the period from about 200,000 to 100,000 years ago.[15] With the Mousterian, Acheulian tools gave way to smaller flake tools, made from a prepared stone core using a technique known as the Levallois technique, and hafting of hand axes was also introduced. From the Acheulian onward, we find clear evidence of planning and design in the manufacture of tools,[16] and as we saw in chapter 5, some archeologists have suggested that this reflects the emergence of language.[17]

Yet the more we compare these accomplishments with the mounting evidence for tool use in

[14] *It's interesting that we can distinguish between voiced and unvoiced consonants, such as /t/ and /d/, even when we whisper. Clearly there is a little more to the distinction than simply the presence or absence of voicing.*

[15] *Schick and Toth 1993.*

[16] *Dominguez-Rodrigo 2001.*

[17] *See, for example, Dennett 1992, Holloway 1969, and Gowlett 1984.*

other species, the less impressive they seem. In the thirty-nine chimpanzee activities listed as showing cultural variation by Andrew Whiten and his colleagues (see chapter 3),[18] the vast majority involve the use of objects that may be described, loosely in some cases, as tools. In some cases the tools are deliberately shaped and show some of the properties of early hominin stone tools. For example, there is a regularity in the shapes of the sticks that chimpanzees use for termiting that is comparable to the regularity of Oldowan stone tools.[19] You will recall how even crows are able to make well-designed tools from pandanus leaves.[20] Moreover, what looks like evidence of design and forethought in hominin stone tools can often be explained as the unintended consequences of toolmaking habits or of the raw material used to make the tools.[21] And even if we do like to think of working in stone as a significant advance, the early toolmakers introduced very little real innovation until the last 300,000 years. One recent commentator remarks, "The Oldowan and Acheulian industrial complexes are remarkable for their slow rate of progress between 2.5 and 0.3 Mya [million years ago] and for limited mobility and regional interaction."[22]

The pace began to pick up a little in the so-called Middle Paleolithic, beginning about 300,000 years ago, both in Africa and in Europe. At this transition point, so-called composite tools became common, with shaped stone pieces mounted in shafts or on handles to form spears and axes. This combining of elements has been likened to language. Technologies show clear regional differences, suggesting the emergence of cultural traditions, which in turn may reflect more sophisticated language.

But the developments of the Middle Paleolithic pale beside the changes that took place in the Upper Paleolithic, beginning some 40,000 years ago. The sudden flowering of technology and art, especially in Europe and Russia, has been termed an "evolutionary explosion."[23] Earlier stone tools of the Acheulian and Mousterian industries were shaped simply by hitting one stone against another to produce flakes, but in the Upper Paleolithic, techniques became more varied. Parallel-sided blades were produced by using a punch that was either held in the hand or impelled by the chest,[24] and shapes began to be formed out of wood, bone, and ivory by cutting and grinding them. Manufactured objects

[18] Whiten, Goodall, McGrew, Nishida, Reynolds, Sugiyama, Tutin, Wrangham, and Boesch 1999.
[19] McGrew, Tutin, and Baldwin 1979.
[20] Hunt 2000.
[21] This point is argued in detail by Davidson and Noble 1993.
[22] Ambrose 2001, 1752.
[23] Pfeiffer 1985.
[24] Fagan 1989.

included projectiles, harpoons, awls, buttons, needles, and orna-
ments.[25] Some of these shapes were of stylized creatures, which
means that they were genuinely designed: the artisans knew how to
make objects according to specifications they had in mind. One example is a half-man half-lion statuette found in southern Germany
and dating from around 30,000 to 33,000 years ago.[26] Widespread evidence exists across Russia, France, and Germany for the weaving of
fibers into clothing, nets, bags, and ropes, dating from around 29,000
years ago.[27]

The remarkable cave drawings discovered in southern Europe
also seem to demonstrate a newfound ability to depict natural objects. They might be regarded as frozen gestures, made possible
by the freeing of the hands from language itself. The cave drawings at Grotte Chauvet in southern France, which date from 32,000
years ago, depict a menagerie of rhinos, bears, lions, and horses.[28]
Although these cave drawings were, until recently, claimed as the
world's oldest, a cave near Verona in Italy has now disclosed even
more ancient drawings (32,000–36,500 years old), of a creature
that is half-human, half-beast.[29] This flowering of art and technology is usually attributed to the arrival of *H. sapiens* in Europe,
which suggests that it reflects earlier developments, probably
in Africa. The archeologist Richard Klein is quoted as saying,
"There was a kind of behavioral revolution [in Africa] 50,000 years
ago. Nobody made art before 50,000 years ago; everybody did
afterwards."[30]

As we saw in chapter 7, molecular evidence now shows that all
non-Africans are descended from a small population that migrated
from Africa only about 52,000 years ago, and it is highly likely that it
was this group of migrants who carried the technology to Europe to
create the evolutionary explosion there. Proba-
bly, though, the development of technology had
been relatively gradual up until that time within
Africa, and its revolutionary effect in Europe was
due to importation rather than on-site invention.
Paul Mellars suggests: "It is possible to point to at
least certain features of the archeological record
of the Middle Stone Age (roughly between

[25] Ambrose 2001.
[26] See Mithen 1996 for this and
other examples of impossible
entities that must have been
figments of the imagination.
[27] Soffer, Adovasio, Illingworth,
Amirkhanov, Praslov, and Street
2000. Textiles are of course
perishable, and the evidence
comes from impressions on clay,
bone, and antlers, and from
fragments of burnt textiles
adhering to stone tools.
[28] Balter 1999.
[29] Balter 2000.
[30] Quoted in Appenzeller 1998,
1451.

100,000 and 40,000 years ago) in Southern Africa which suggest a significantly more 'complex' (and perhaps more 'advanced') pattern of behavior than that reflected in the parallel records of the Middle Paleolithic in northern Eurasia over the same time range."[31] For example, a bone industry, which probably included the manufacture of harpoons to catch fish, has been discovered in the Republic of the Congo, and dates from about 90,000 years ago.[32] It has been also been suggested that the first colonization of Australia, over 60,000 years ago, is earlier evidence of modern human behavior, including language, partly on the grounds that it would have required the use of seagoing craft, capable in one case of crossing a distance of ninety kilometers.[33] However, it now appears from the mitochondrial witness of Mungo Man that these early rafters may not be the direct forebears of modern humans; present-day indigenous Australians may be the descendents of a later wave of immigrants.

Some authors have nevertheless argued that the emergence of art and manufacture around 50,000 years ago took place so suddenly that it calls for some special explanation. Richard Klein is quoted as saying that it was "a biological advance."[34] Others have proposed that it was language itself, and that cave art reflects the dawn of symbolic understanding.[35] It seems unlikely, though, that such a complex function as language could have emerged as a biological entity in so short a period of time, although linguists such as Derek Bickerton have argued that syntax might indeed have depended on a fortunate mutation— what has been called the "big bang" theory.[36] Yet others have argued that art and technology were simply cultural inventions, handed down from generation to generation.[37]

My own suggestion is that the final achievement of autonomous speech freed the hands and opened up the full potential for manufacture, ped-

[31] *Mellars 1989, 367.*
[32] *Yellen, Brooks, Cornelissen, Mehlman, and Stewart 1995.*
[33] *Davidson and Noble 1992.*
[34] *Quoted in Appenzeller 1998, 1452.*
[35] *See, for example, Noble and Davidson 1996.*
[36] *Bickerton's 1995 argument is regarded as tantamount to creationism by some critics. Bickerton argues, though, that the essential elements of grammar were already present, and all it took was the equivalent of the throw of a switch to get them up and running. Here's how he put it: "Imagine a newly constructed factory lying idle because someone neglected to make a crucial connection in the electrical wiring. The making of that single connection is all that is needed to turn a dark and silent edifice into a pulsating, brilliantly lit work place. This could have been how syntax evolved in the brain" (83). I must say I find it hard to believe that evolution could work this way. More recently, though, Bickerton seems to have modified his view, arguing that syntax may have been gradually refined through a process of Baldwinian evolution (Calvin and Bickerton 1999).*
[37] *Randall White is quoted as saying, "I think that what we call art is an invention, like agriculture, which was an invention by people who were capable of it tens of millennia before" (quoted in Appenzeller 1998, 1452).*

agogy, and cultural transmission of information.[38] But this achievement is unlikely to have depended on a sudden biological change. Rather, the adaptations necessary for autonomous speech were probably in place 100,000 years earlier, with the emergence of *H. sapiens* in Africa. Vocalization must have played a prominent role in language even then, for otherwise the biological adaptations necessary to produce articulate sounds would scarcely have evolved. Nevertheless, my guess is that language still depended in part on manual and facial gestures as well as on vocal accompaniment, perhaps until as recently as 50,000 years ago.

If this scenario is correct, then autonomous speech must have been an *invention* rather than the immediate outcome of some anatomical change—an invention as powerful in its way as the invention of other forms of oppression, such as the gun, or the Internet.

■ The *invention* of autonomous speech ■

The idea that speech may have been an invention was anticipated by none other than Charles Darwin: "Man not only uses inarticulate cries, gestures and expressions, but has invented articulate language; if, indeed, the word *invented* can be applied to a process, completed by innumerable steps, half-consciously made."[39] It may indeed sound strange to refer to autonomous speech as an invention when it seems so natural and universal, but remember that signed language seems equally natural to those who use it from an early age. Some authors, such as Andrew Lock, have gone so far as to suggest that language itself is an invention, structured by social relations, and making use of available biological capacities.[40] Indeed, he has insightfully referred to the development of language in children as "the guided reinvention of language."[41] In Lock's view, contrary to the claims of Pinker, language is not a biological instinct but

[38] Not everyone agrees that the advance of technology can be attributed to the freeing of the hands. Mithen 1996, for example, objects on the grounds that technology prior to the evolutionary explosion involved as much "manual dexterity" as that which followed. But the switch to autonomous vocal language would scarcely have affected manual dexterity. Gestural language itself would have required considerable dexterity, in order to create enough distinct hand shapes and arm movements to provide a sizeable vocabulary. Acheulian technology implies considerable dexterity, as Mithen points out, but was limited in complexity and variety. With a switch to autonomous vocal language, language and manufacture could take place in parallel, and language could play a supportive rather than a competitive role. If fully autonomous vocal language did not emerge until the Upper Paleolithic, this could be precisely why the widespread manufacture of what Mithen calls "multi-component tools" did not appear until then. The general effect of separating language from manufacture might actually have been to decrease manual dexterity, as manufacture came to depend more on the combining of components and sequential planning. The more we depend on machines, the less room there is for the craftsman's touch.

[39] Darwin 1965, 60.

[40] Lock 1999.

[41] This is in fact the title of his 1980 book on language development in children.

rather a socially constructed enterprise that conforms to our biological makeup.

My own view, though, is less radical, and more Pinkerian. I suspect that language itself is very largely a matter of biological adaptation achieved through natural selection, but perhaps encompassing abilities broader than just communication, such as the ability to take the mental perspectives of others.[42] But the idea that we might use vocalization to refer to objects and actions, as well as convey syntax, thereby eventually rendering manual gestures largely (but not completely) superfluous, may well have been an invention, carried on through social custom from generation to generation. Of course the adaptations of the vocal apparatus and cerebral mechanisms to permit articulate speech were fundamentally biological and must have emerged gradually, probably over the past 2 million years or so, and may have been complete by the time *Homo sapiens* came on the scene, some 170,000 years ago. But then, even as now, though to a greater extent, language was probably an amalgam of gesture and vocalization. The discovery that language could be *autonomously* vocal might have yielded a rapid technological payoff: more complex technologies could be described, explained, and transmitted between generations.

The invention of autonomous speech may have been as much a matter of eliminating manual gestures as of inventing new spoken words. Speech may well have emerged incrementally. For example, it may have first appeared as grunts accompanying gestures; then the range of oral gestures was expanded by making them audible, or voicing was added to oral sounds to create new variants, such as /d/ from /t/ and /b/ from /p/. But as this tendency grew, so the need for manual gestures declined, and the realization that language might be conveyed *entirely* through sound may have been essentially a cultural one. Proto-World may therefore have been the invention of a small group of *Homo sapiens* who were subsequently to conquer the world.

We humans have manifestly invented a number of other complex skills, such as playing the piano or tennis or the stock market, constructing tools, juggling, and typing. All, of course, depend on biological preadaptations, and it is a safe bet that other species will not be able to match these accomplishments. By

[42] *A computer programmer in the department where I work is a recent immigrant and may well represent the next stage in hominin evolution. Not only does he have instant rapport with my computer but when I phone him for help, he answers, "It's Henry there." Already he has taken the mental perspective of the distraught caller.*

the same token, however, we are unlikely to match the ability of fish to swim, or of seals to throw and catch balls with their noses, or of birds to fly—although we can compensate through technology, as illustrated by that monstrous bird, the jumbo jet. There is one accomplishment that is especially useful to compare with speaking, because it too has to do with language. That accomplishment is writing—and of course its complement, reading.

■ Lessons from writing ■

There can be no doubt that speech is more "natural" than writing. As Darwin put it, "[Speech] differs . . . from all the ordinary arts, for man has an instinctive tendency to speak, as we see in the babble of our young children; whilst no child has an instinctive tendency to brew, bake, or write."[43] Speech is universal, at least among the hearing population, whereas until quite recently writing and reading were accomplishments restricted to a privileged minority. It has been estimated that even in the United States, some 10 to 20 percent of the population are functionally illiterate, and in some African countries the proportion may be well over 50 percent.[44] Everyone knows, too, that learning to read and write is hard work, whereas learning to talk and understand speech is so easy and natural that we can't even remember when it happened.

Yet once we have learned to read and write, these skills become as automatic as speaking and understanding. Just as one hears a spoken word as a word, and not as a jumble of sounds, so one sees a printed word as a word, and not as a bunch of squiggles. Both spoken and written language depend on brain regions in the left side of the brain, at least for most of us. Of course, part of the difference between vocal and written language is that writing, at least in our culture, is based on the prior acquisition of speech, and one of the reasons that learning to read and write is so much more difficult than learning to speak may have to do simply with difficulties in mapping printed words onto spoken ones, rather than with processing print per se. Other forms of communicating by inscribing may well be more natural and immediate, as we shall see below. So let's take a closer look at the history of writing, and see what it might tell us about the evolution of speech.

[43] Darwin 1896, 87.
[44] Crystal 1997. Illiteracy is hard to define with precision, though, since some cultures, such as the Pueblo, referred to below, have developed systems of communicating that lie somewhere between writing and art.

The origins of writing might perhaps be traced to the cave drawings of 30,000 to 40,000 years ago. Indeed, as suggested above, these might even have been an indirect consequence of the invention of autonomous vocal speech, which freed the hands for doodling. But these pictures did not begin to assume languagelike properties until they developed standard forms known as *pictograms*, designed expressly for the purpose of visual communication. The earliest known pictograms date from Sumeria in Mesopotamia and spread to surrounding areas around five thousand years ago. Similar systems were developed independently in other parts of the world, including China and South America. Pictograms later evolved into *ideograms*, in which the symbols stood for abstract as well as concrete ideas, and later still these evolved into *logograms*. Logograms stand for morphemes, the basic units of meaning, thus moving writing closer to speech itself.

One system that is of special relevance to the theme of this book is American Indian pictography, more popularly known as rock art but perhaps better called rock writing, which is based on Indian signed language rather than on speech. Carol Patterson-Rudolph has extensively studied the shapes, known as petroglyphs, carved into the rocks in various locations in North and South America, and has uncovered many of the symbols, metaphors, and rules that govern their formation. Some of the shapes are stylized pictures of people or animals; others are conventionalized symbols for actions or concepts; yet others are stylized representations of the handshapes used in signed language.

In rock writing, pictures of animals are used as metaphors. For example, the mountain lion is a great hunter, so a picture of a lion evokes the concept of hunting with the power and skill of a lion. The roadrunner stands for courage and protection against enemies. Pictures of animals are, of course, a feature of the earliest cave paintings found in Europe and may have been used metaphorically there too. Pictures of animal tracks may also have specific meanings. In petroglyphs made by the Pueblo in the Rio Grande, for example, turkey tracks frequently appear. Turkeys are distinguished from other birds in that they have only three toes, so pictures of the tracks are sufficient to identify the species. The Pueblo know that turkeys are generous by nature, and that they depend on the older males to lead them to food and water. Turkey tracks, therefore, indicate the direction to something of general interest, which can be identified from other symbols in the panel. What is intriguing about these systems of petroglyphs is that they tell narrative stories through pictures and abstract signs, without any direct appeal to speech at all. This is a kind of visual language, with its own syntax and symbols,[45] and may well be a more natural and accessible form of inscribed language than conventional writing. And it is an invention, not a biological given.

In other systems of writing, logograms have typically become associated with speech rather than gesture and have tended to lose their iconic aspect—a further example of the process of conventionalization. In Chinese and in Japanese Kanji scripts, logograms have survived that have only peripheral references to speech sounds, but in other scripts written symbols were connected more tightly with speech sounds and eventually lost any but the most rudimentary pictorial association. One kind of speech-based script is the *syllabary*, in which each symbol stands for a syllable and conveys no meaning by itself. The first syllabary may have been invented by the Semites and Phoenicians around 1700 B.C.; later it was adapted in Old Hebrew, Cypriote, and Persian scripts. Two surviving syllabaries are the Japanese Katakana and Hiragana scripts. But syllabaries have largely given way to alphabetic scripts, in which the individual symbols, or letters, stand for phonemes.[46] Alphabetic scripts have the virtue of economy, since relatively few symbols are required to represent tens of thousands of words.[47]

The point to be drawn from this is that we humans no doubt had the potential for writing for tens of thousands of years before writing systems were actually invented, and there are still large numbers of people who cannot read or write. The biological prerequisites for writing include manual dexterity, spatial sense, and language itself—and speech would also have been a prerequisite for writing systems that map onto speech sounds. Similarly, we may have possessed the biological prerequisites for spoken language well before autonomous spoken languages were invented. If my analysis is correct, the evolution of writing may also share another feature with the evolution of language: it has *con-*

[45] Patterson-Rudolph 1993. Patterson-Rudolph's fine work follows in the tradition of the eminent archeologist and ethnologist LaVan Martineau, whose pioneering book The Rocks Begin to Speak drew attention to the sophistication and eloquence of rock art.

[46] Technically, the symbols are known as graphemes rather than letters, and the mapping between graphemes and phonemes is not perfect. For example, some phonemes are represented as combinations of letters, as in th, ch, sh, ph, and ng, and some letters, such as c, g, and y, do double duty.

[47] The foregoing is based largely on the classic work of Gelb 1974, but see also the more recent studies by Gaur 1987 and Coulmas 1989.

ventionalized, evolving from highly iconic pictorial representations to abstract graphemes, just as language itself evolved from iconic gestures to abstract, spoken words. As we have seen, the progression from iconic to abstract may be a natural one in the evolution of any communication system.

There can be no question that the invention of writing has had a profound influence on human life, greatly improving the transmission and accumulation of technology and cultural practices. Indeed, it may not be an exaggeration to suggest that societies that have remained illiterate, or largely so, are themselves in danger of serious attrition, if not extinction, not so much through the wielding of force by societies with more advanced technologies as through poverty, famine, and disease. It may well be that the invention of autonomous speech had a similar impact 50,000 years ago, leading eventually to the extinction of *Homo erectus,* the Neanderthals, and even those earlier migrating members of our own species, such as Mungo Man, who still relied in part on manual gestures. If this is so, then our masculine forebears may not have been quite the warlike rapists depicted earlier. Those without speech may have died of natural causes; they were not necessarily murdered by our talkative ancestors.

But while this scenario may paint us in a slightly more attractive light, it might also serve as a warning. Survival of humans societies may well depend on the development of increasingly advanced technology and communications systems; yet there are still large numbers of people in the Third World who are not even literate, let alone numerate or scientifically aware. Can we reverse these imbalances, or must we repeat what happened 30,000 years ago, when those technological, artistic, and, I submit, garrulous members of ours species displaced everyone else?

But Language *Is Biological*

If autonomous speech is indeed an invention, this does not, of course, mean that it has no biological component. We saw in chapter 7 that the vocal tract and the cerebral control of voicing and articulation had to change considerably to make speech possible, and these changes must have been driven by the adaptive advantages of adding vocalization to the language repertoire. I think it is safe to say that chimpanzees will not invent speech—or if they do, it will take a couple of million years of preadaptation. Further, language itself clearly has a strong biological component. The biological requisites for recursive grammar probably began to emerge some 2 million years ago, perhaps first in the context of the kinds of recursion necessary for higher-order "theory of mind." Another requisite, at least for producing or understanding complex sentences, would be an enhanced short-term memory capacity, so that several levels of recursion can be held in mind. It may have been this requirement that underlay the increase in brain size that has occurred over the past 2 million years.

But I suspect that grammar, with its recursive structure, was developed first in the context of gesture, with the vocal element emerging later, with modifications to the vocal tract and the developing cortical control of vocalization and breathing. The readiness with which deaf children learn and even improvise signed language shows that it is just as "natural" as speech, and perhaps even more so. If so, language in the predecessors of *Homo sapiens*, and perhaps even in early *Homo sapiens*, may still have been fundamentally gestural rather than vocal.

■ Grammar and the generative mind ■

I have suggested, then, that grammatical language evolved, primarily as a gestural system, starting about 2 million years ago, when brain size began to expand and our *Homo* ancestors began to migrate out of Africa. Precisely when the conversion from protolanguage to grammatical language took place is difficult to determine, although it was probably not a sudden event. The development of technology probably does not provide useful clues, since it proceeded only very slowly from the Oldowan and Acheulian cultures until the evolutionary explosion some 40,000 years ago. This explosion, I have argued, reflects the invention of autonomous *speech*, and earlier technological development may in fact have been hampered by the competing role of gesture.

Nevertheless, the way we manipulate mechanical objects may give some clues as to how grammar evolved, if not when. Patricia Greenfield has given an account of how children simultaneously develop hierarchical representations for both language and the manipulation of objects.[48] Just as they begin to combine words into phrases

---

[48] *Greenfield 1991.*

and then phrases into sentences, so they begin to combine objects, such as nuts and bolts, or nesting cups, and then use the combinations as objects for further manipulation. Greenfield argues that both activities depend on Broca's area on the left side of the brain. She goes on to suggest that this relation between language and hierarchical manipulation persists into adulthood, citing evidence that people with Broca's aphasia (a deficit in speech following damage to Broca's area) are also poor at reproducing drawings of hierarchical tree structures composed of lines.[49]

However she also cites evidence that, in a sample of mentally retarded children, some were skilled in hierarchical construction but deficient in grammar, while others showed the reverse pattern. She relates these findings to neurophysiological findings that up to the age of two, the same brain area may be involved equally in both functions, but beyond that age there is increasing differentiation of Broca's area: an upper region organizes the manipulation of manual objects, and an adjacent lower region, grammar. In many cases of brain injury, both regions will be damaged, resulting in combined deficits, but in some cases one or other region might be damaged, resulting in deficits either of manipulation or of grammar.

Our ability to construct objects indeed has many of the properties of language, including a capacity to generate an infinite variety of complex structures, in hierarchical fashion. We combine elements of construction in much the same way that we combine phonemes to form words, words to form phrases, phrases to form sentences. The "phonemes" of construction include bricks, boards, nails, nuts, bolts, screws, brackets, handles, wheels, axles, hinges, and so forth. At a higher level in the hierarchy we have tables, chairs, doors, engines, central processing units, and so forth. And then houses, buildings, automobiles, ships, airplanes, motorized golf carts, computers—well, you name it. Irving Biederman has also suggested that we represent and recognize objects in our minds as combinations of standardized shapes that he calls *geons*, a kind of geometric analogy of phonemes.[50]

For clues as to the recursive aspect of language, we might look to so-called "theory of mind," which is the ability to understand what is going on in the minds of others. This is recursive, because our understanding includes the fact that we can also understand that others can

[49] *Grossman 1980.*
[50] *Biederman 1987.*

understand what is going on in the minds of (yet) others—and so on. As I pointed out in chapter 5, recursive sentences such as *I suspect that she knows that I'm watching her talking to him* express correspondingly recursive thoughts, and indeed, there is evidence that children develop theory of mind at around the same time that they develop recursive syntax. Again, though, the two can be dissociated. For example, autistic children appear to be deficient in theory of mind but do develop syntactic language, although they are typically lacking in the pragmatic aspects of language, such as the ability to understand irony or decipher the speaker's intentions by "reading between the lines."[51] Nevertheless, theory of mind seems to depend on the frontal lobes,[52] and perhaps even on those ever-obliging mirror neurons.[53]

I have elsewhere referred to a *generative assembling device*, or GAD, which may operate in a variety of contexts, including language, to create the enormous variety of structures that populate our worlds.[54] Music, of course, also has a generative structure (will we ever run out of tunes?). This does not mean that all of these varied activities depend on precisely the same set of generative rules. A common core may underlie them, but as Greenfield suggests, there has probably been differentiation of skills for different purposes. This may have depended in turn on the differentiation of neural tissue in and around Broca's area in the frontal lobes.

My guess is that this differentiation comes about as a function of growth. The actual patterns of representation would depend in part on biologically programmed patterns of growth, but also on the nature of experience during periods of growth. In particular, the left side of the brain seems to undergo a growth spurt roughly between the ages of two and four,[55] and it is during this period that grammar becomes embedded in the child's brain. During this period, children are typically confined to play settings where they are exposed to language and to toys, so that the constructive skills of language, object manipulation, and theory of mind are acquired. The child who is also heavily exposed to music and musical training at such an early age may well become another Mozart, impregnated with the *grammar* of music. There is evidence that, in professional musicians trained from an early age, the areas involved in the various aspects of musical performance, including the ability to read music, correspond quite closely to the

[51] *Baron-Cohen 1999.*
[52] *Stone, Baron-Cohen, and Knight 1999; Stuss, Gallup, and Alexander 2001.*
[53] *Gallese and Goldman 1998.*
[54] *Corballis 1991.*
[55] *Thatcher, Walker, and Guidice 1987.*

areas involved in language. These areas are on the left side of the brain and include one that is close to Broca's area.[56]

But a later growth spurt in the right brain may capture other complex skills on that side of the brain. Even if we are not blessed with intensive musical experience early in life, Broca's area may play a role in learning what might be termed the syntax of music.[57] For us musical slobs, however, there is, if anything, a slight dominance of the equivalent of Broca's area on the *right*, which might come as a relief to its left-sided twin, with its already crowded duties. This later period of right-sided growth might also coincide with increased mobility and exposure to the spatial environment and explain why spatial abilities also tend to be right-brained.[58]

In chapter 1, I described the dispute over the question of whether grammar is innate, as proposed by Chomsky, Pinker, and others, or whether it can be learned by a connectionist system that grows as it learns, as Jeff Elman argued. Both views may contain elements of truth. The genes may program patterns of growth that allow different complex, hierarchical skills to be acquired, in part by organizing different rates of growth on the two sides of the brain. But differentiation may occur even within one or the other side. Consider Broca's area, for example, which has been implicated in the programming of gesture, speech, theory of mind, and music! I doubt that it is really so versatile. There are probably subareas in the vicinity of Broca's area (which is itself, in fact, poorly defined) that become involved with different abilities. Perhaps the development of orchestrating systems in the frontal lobes may be likened to the growth of a flower, with the petals becoming differentiated over time, each responsible for increasingly distinct abilities.

But we must not lose sight of the flower itself, which captures something of the nature and unity of the human mind. Alan Ayckbourne, the English dramatist, may have understood the dangers of focusing too much on the petals, when he had one of the characters in his play *Table Manners* say, "If you gave Ruth a rose, she'd peel all the petals off to make sure there weren't any greenfly. And when she'd done that, she'd turn round and say, do you call that a rose? Look at it, it's all in bits."[59]

The idea that some general principles might govern a variety of generative abilities helps ex-

[56] *Sergent, Zuck, Terriah, and MacDonald 1992.*

[57] *Maess, Koelsch, Gunter, and Friederici 2001.*

[58] *This theme is explored in more detail in chapter 11 of my book* The Lopsided Ape *1991.*

[59] *From act 1, scene 1, of* Table Manners *1975.*

plain why grammar is amodal, as much at home in gesture as in speech. If the same basic principles apply to the structuring of complex manufacture, theory of mind, and even music, then they can surely embrace both signed and spoken language—and even the transitional combination of the two. The interaction of programmed growth with experience may provide for hierarchical structure and recursion, but the details of structure will then depend on other constraints. For example, some of the grammatical properties we associate with speech have to do with *linearization*: the requirement that our descriptions of a four-dimensional world of space and time be squeezed into the single dimension of time. Many of the rules of syntax, at least, have to do with ordering and with the shifting of elements from one location to another. Rules about word order, for example, govern the construction of the so-called *wh-*questions in English, questions involving *what, where, why,* or *when,* such as are frequently asked by four-year-olds.

Take a sentence like *She put the ice cream in the fridge.* One might then ask, *Where did she put the ice cream?* or *What did she put in the fridge?* or, especially if you're a four-year-old, *Why did she put the ice cream in the fridge?* In each case the couplet *Wh- did* is introduced at the beginning, and in the first two examples the critical phrases *in the fridge* and *the ice cream* are deleted from the original. In signed languages, however, information is not so rigidly constrained into a linear sequence, and different aspects of the message can be transmitted in parallel. A statement can be turned into a question without any rearrangement of its components if accompanied by a change in posture and a raising of the eyebrows, as I explained in chapter 6.

■ So when did grammatical language evolve? ■

If the emergence of grammatical language from protolanguage does indeed depend on programmed patterns of growth, then it is reasonable to suppose that the process began with the increase in brain size relative to body size, some 2 million years ago. But perhaps it wasn't increased brain size per se that did the trick but rather the prolongation of childhood, the fact that most of the growth of our large brains takes place after birth. It was this oddity that allowed environmental influences to interact with growth, and so embed hierarchically organized structures in the brain. As we saw in chapter 5, this characteristic was probably in place in *Homo erectus* around 1.6 million years ago. Perhaps a form of recursive grammar had also evolved by this stage.

Another estimate comes, interestingly if unexpectedly, from an analysis of grooming. I am not referring to the time spent by vain members of our species in front of the mirror, but rather to the habit many primates have of picking through each other's fur to remove fleas and small pieces of debris. The anthropologist Robin Dunbar has argued that grooming is a precursor to language.[60] This may seem a strained argument—is Dunbar suggesting that language is merely a compensation for the loss of body hair, a way of picking the nits out of each other's minds, now that our bodies no longer provide good foraging? Nevertheless, he may have a point. Grooming is certainly a form of social communication; it implies some degree of ability to take the mental perspective of others and can even be considered an example of reciprocal altruism, along the lines of "you scratch my back and I'll scratch yours." Dunbar points to the importance in human society of gossip, which he considers a form of grooming.

Dunbar has also shown that the time spent grooming is related to two other characteristics of primates, one social and one neurological. Primates tend to form themselves into groups, partly as a means of defense against predators (there is safety in numbers). The size of the group varies with the species and tends to increase with the ratio of the size of the neocortex to the rest of the brain, the so-called *neocortical ratio*. A notable exception to this rule, though, is the rather solitary orangutan.[61] Humans have the largest neocortical ratio, at 4.1, clearly ahead of our close relative the chimpanzee, at 3.2. Gorillas weigh in at 2.65, orangutans at 2.99, and gibbons at 2.08. According to Dunbar's equation relating group size to neocortical ratio, humans should belong to groups of 148, give or take about 50. This appears to be reasonably consistent with estimated sizes of early Neolithic villages. Modern cities confuse matters, but maybe if you add together your old school chums, work mates, soccer team, friendly neighbors—oh, and family, perhaps excluding the odd wayward uncle—then it might add up to a figure like this. Count the people at-

[60] See Dunbar 1993.
[61] See Byrne 1995 for further discussion. Byrne also shows that neocortical ratio is positively related to the prevalence of tactical deception in various primate species. Tactical deception is another index of so-called "theory of mind."

■ Table 9.1 ■

Predicted amount of time spent grooming, based on neocortical ratio (data from Dunbar and Aiello 1993)

Species	Percent of time grooming
Australopithecus/Praeanthropus	18.44
Homo habilis/rudolfensis	22.73
Homo erectus/ergaster	30.97
Early *Homo sapiens* (or *H. heidelbergensis*)	37.88
Neanderthals	40.66
Modern *H. sapiens* (male)	37.33
Modern *H. sapiens* (female)	40.55

tending the next wedding and funeral you go to, and take the average.[62]

Dunbar's magic equations can use group size to predict the percent of time spent grooming, and the predictions for various hominin species are shown in table 9.1 (and you thought that it was females who gossiped!).[63] Dunbar argues that with the increase in time spent grooming and the increase in neocortical ratio, there comes a point when a less time-consuming form of grooming is required. You can't spend too much of the day grooming, although I could name people who seem to spend most of their time gossiping.[64] In any event, Dunbar suggests that the critical point is 30 percent. As the table shows, this suggests that language, presumably in the form of gossip, may have emerged with late *Homo erectus* or early *Homo sapiens*.[65]

This may be as good a guess as any, although a slightly later date would coincide with the beginning of the Middle Paleolithic, around 300,000 years ago. As we saw earlier, composite tools became widespread from this time, perhaps indicating the emergence of generative grammar. Other indicators suggest an earlier date. One is the controlled use of fire, which is a uniquely human activity that may date from 1 to 1.5 million years ago, although this is debated.[66] Controlling fire requires group consensus, cooperation, and planning, all of which imply effective communication.[67] Evidence of burials has been taken to imply an understanding of death and perhaps religious belief, which

[62] *The funeral crowd is likely to yield an underestimate, since some of the other clan members will probably have already died. On the other hand, weddings will overestimate, since two clans usually attend.*

[63] *Dunbar and Aiello 1993.*

[64] *To actually name them would be to indulge in gossip myself, and most distasteful.*

[65] *The transition from grooming to gossip is about as simple a version of the gestural theory as one could hope for, although I'm sure Dunbar didn't mean it quite this way.*

[66] *Brain 1993.*

[67] *Ronen 1998.*

might again imply language. Although the evidence goes back only perhaps 100,000 years, there are indications that the Neanderthals, as well as *Homo sapiens*, ritually buried their dead, which suggests that some sort of recognition of death may go back to the common ancestor, perhaps 500,000 years ago. In any event, a syntactic gestural language, comparable to modern signed language but perhaps including some vocal elements also, may well have emerged by then.

■ Conclusions ■

The main theme of this chapter, and the last plank in my argument for the gestural origins of language, is that it was not the emergence of language itself that gave rise to the evolutionary explosion that has made our lives so different from our near relatives, the great apes. Rather, it was the invention of autonomous speech, freeing the hands for more sophisticated manufacture and allowing language to disengage from other manual activities, so that people could communicate while changing the baby's diaper, and even explain to a novice what they were doing. The idea that language may have evolved relatively slowly, with grammar beginning to form as early as 2 million years ago, seems much more in accord with biological reality than the notion of a linguistic "big bang" within the past 200,000 years. Language and manufacture also allowed cultural transmission to become the dominant mode of inheritance in human life. That ungainly bird, the jumbo jet, could not have been created without hundreds, perhaps thousands, of years of cultural evolution, and the brains that created it were not biologically superior to brains that existed 100,000 years ago in Africa.

One of the challenges in piecing together an understanding of how our own species evolved is to explain the gap between the emergence of *Homo sapiens* 170,000 years ago and the appearance a mere 50,000 years ago of that dominating, technologically sophisticated cohort that eventually populated the globe. Clearly, these people had something new going for them, and there is no evidence that the secret of their success lay in their biology. Nor is it likely that they suddenly invented *language*. What they had done, I think, was to eventually rid language of the necessity to use gesture, with enormous consequences for manufacture, art, ritual, and culture generally.

I have argued that the emergence of fully autonomous speech may have been an invention rather than a fait accompli of biology. After all, many of the subsequent developments in manufacture have depended on the invention of other means of communication, such as writing, mathematics, and computing. The invention of speech may have been merely the first of many such developments that have put us not only on the map, but all over it.

10 ■ Synopsis ■

Birds do it; bees do it; even educated Australians do it. They *communicate*. Birds sing to establish territory or emit cries to warn of danger or to tell the world that it's spring. Bees indulge in a characteristic "waggle dance" to indicate to their fellow bees the location of food. Fireflies exchange light flashes to indicate sexual readiness.[1] Darwin was impressed by the "considerable powers of communication" shown by ants, using their antennae.[2] A peculiarly solipsistic form of communication occurs in bats, who effectively talk to themselves by means of their sonar system of echolocation, enabling them to navigate through dark caves and, no doubt, belfries, and to hunt for prey in forests at night. In this system, the animal emits pulses of sounds and computes the location of objects from the ensuing echo, and is thus both sender and receiver of the message.[3] But bats also communicate vocally with other bats, using echolocation signals from others to help find prey.[4]

In vertebrate evolution, as we have seen, the use of vocalization as a means of communication may go back to the anurans, predecessors of frogs, some 170 million years ago.[5] The most conspicuous call in the frog's repertoire is the delightfully

[1] *These and many other examples are given in Griffin 1976.*
[2] *Darwin 1896, 89.*
[3] *Hauser 1996.*
[4] *Balcombe and Fenton 1988. It appears that bats do not intentionally signal the presence of prey to other bats but simply adventitiously pick up echolocation signals from them. This rather suggests that bats do not possess theory of mind. I'm sorry you had to wait this long to learn about bats.*
[5] *Ever wondered about that frog in your throat? Now you know where it came from.*

named advertisement call, which is used by male frogs both to attract mates and to put male competitors on notice. As in humans, the male frog's larynx is larger and heavier than that of the female, giving him a deeper voice.[6] Vocalization in present-day frogs appears to be largely under left-hemispheric control, as it is in birds, rodents, and primates, including ourselves, suggesting that cerebral asymmetry for vocalization may go back to our common vertebrate ancestor, maybe 170 million years. Language is a capacity we like to jealously claim as our own, but it would be perverse to think that human speech did not make use of earlier adaptations to the vocal cords and brain underlying the production of sound in vertebrates generally.

Of course, this is not to say that these other species possess true language; their vocalizations are largely tied to instinctive, emotional situations like bonding, mating, territory claiming, and sounding alarms. In understanding the evolution of language itself, we need to consider the emergence of behaviors that might be described as voluntary, allowing for a level of improvisation that would have greatly expanded the range of possible communication and freed it from the slow grip of natural selection. Following the destruction of the dinosaurs around 65 million years ago, mammals came into their own, including a primate known as *Purgatorius*. The primates gave us forward-placed eyes and stereoscopic vision, color vision, and grasping hands. These characteristics led naturally to forms of behavior and expression that involve bodily movement rather than the production of sound, and to a dominance of vision and touch over audition.

The discovery of "mirror neurons" in the monkey brain provided a significant boost to the notion that language may have originated in gestures, since these neurons respond both when the monkey makes a grasping movement and when it observes the same movement made by others. This is the kind of mapping that one might expect to find in a sophisticated communication system, where sender and receiver must share the same understanding. Since the mirror-neuron system is present in both monkeys and humans, it was most likely present in the common ancestor, presumably before the split between apes and monkeys over 30 million years ago. As we saw in chapter 3, this system maps the production of specific reaching and grasping movements onto the perception of those same movements made by another individual: monkey see, monkey do. Same goes for me: as I watch a rugby game, I find myself squirming and wriggling in my

[6] *Hauser 1996.*

chair in helpless and ineffective concert with the muddied oafs on the TV screen. This system provided a basis for a form of communication that was voluntary and flexible rather than fixed, as in bird calls.

Something over 30 million years ago, we went ape, differentiating ourselves from the Old World monkeys. By around 16 million years ago, the larger-brained great apes had split off and are now made up of orangutans, gorillas, chimpanzees, bonobos, and humans. Larger brains probably heralded an increase in what might be termed "off-line" thinking, including enhanced representation of objects in the brain, so that problem solving can be accomplished mentally rather than through physical trial and error. The great apes are also capable of learning protolanguage—the combining of objects and actions to make simple requests—although there is little evidence of protolanguage among great apes in the wild, present company excepted. Nevertheless the great apes gesture extensively in the wild, and there seems little doubt that voluntary activity, involving intentional acts and advance planning, evolved primarily in the context of movements of the limbs: eating, moving about, grooming, using tools. The overt behavior of great apes provides a platform for language, even though it does not constitute language itself, and they do not seem to have intentional control of vocalization or the flexibility of vocal and oral programming necessary to produce anything approaching speech. Some have argued that the great apes, but not other primates, have the rudiments of "theory of mind," as evidenced by such behaviors as tactical deception and self-recognition, although laboratory studies have been equivocal at best.

Around 5 or 6 million years ago, we stood up. Bipedalism was the main characteristic of the hominins that distinguished them from the other great apes. No one knows why our enterprising ancestors did this. Maybe they were just upstanding folk. The notion that it was an adaptation to savannalike conditions seems to have gone out of style. One possibility is that the great apes of East Africa were forced more and more into forested environments that bordered on lakes, rivers, or the sea and took to foraging in the water; that is, bipedalism may have been an adaptation to wading. However it happened, bipedalism would have freed the hands and arms for more effective gesturing.

But the advance from protolanguage to true grammatical language may not have begun until something over 2 million years ago, when the genus *Homo* emerged. Indeed, it may have been at this point that some members of this genus, or at least those who remained in Africa, were forced onto a more savannalike terrain. This branch of the hominins was distinguished by an increase in brain size, the invention of stone tools, and starting about 2 million years ago, the beginnings of multiple migrations out of Africa. It is likely that language became increasingly sophisticated from then on. This may have been driven in part by the development of recursive thought, in which structures can be nested one within another. This includes the embedding of concepts such as seeing or knowing, as in *I know that he is seeing her.*

This kind of embedding may characterize some of our more complex thoughts, as well as the language used to communicate them. Nonhuman great apes may be capable of first-order recursion—for example, seeing that they are being seen—but may use it only in certain situations, such as seeing or hearing or being afraid. What may distinguish human thought is the understanding of recursion as a general principle, so that embedding can go beyond first-order recursion (*I see that he sees that I see him*) and apply to a wider variety of situations. Recursive thought may have been driven in the first instance more by social pressures, and the subtle calculus of competition and cooperation, than by the requirements of language itself.

Recursive thought may also underlie mental time travel, which some have claimed as a uniquely human faculty.[7] We can, for example, understand that yesterday we thought it was going to be fine today, whereas in fact the picnic was ruined by rain. Recursion is perhaps implied in any act of transporting ourselves mentally from our own location in the present to another time or another spatial location or another person's point of view. We humans seem to be rather good at escaping current reality in these ways. Recursion is also required for the language used to describe these mental acts.

These enhanced mental and linguistic capacities may also underlie the early migrations out of Africa. These migrations were presumably not seasonal, driven by genetically controlled hormonal mechanisms, as in the case of migratory birds or animal herds, but involved moving into unknown territory and adapting to new conditions. Such changes in the lifestyle of the group probably required planning and reference to past experience as well as future expectations.

For most of this period, I have argued, language would have

[7] *Suddendorf and Corballis 1997;
Wheeler, Stuss, and Tulving 1993.*

been primarily gestural, although increasingly punctuated by vocalization. Articulate speech would have required extensive changes to the vocal tract and the cortical control of vocalization and breathing, and the evidence suggests that these were not complete until relatively late in the evolution of the genus *Homo*. Indeed, they may not have been complete even in the Neanderthals of 35,000 years ago, although this claim is controversial. The adaptations necessary for articulate vocalization may have been selected, not as a replacement for manual gestures, but rather to augment them. Some gestures were no doubt facial, as are some of the signs of present-day sign language, and vocalization may have served in part to augment facial and oral gestures and to make invisible gestures of the tongue and oral cavity audible. Even today, of course, language is scarcely ever purely vocal. Watch any two people conversing, and you will see that their speech is accompanied by manual gestures and facial expressions that enhance meaning.

And, of course, the signed languages of the deaf have shown us that articulate, grammatical language *can* be carried on entirely with gestures of the hands and face, in the absence of any sound whatsoever.

Many species show a left-hemispheric dominance for vocalization—a bias that may go back to the very origins of the vocal chords, perhaps 170 million years ago. As vocalizations were increasingly incorporated into manual gesture, this may have created a left-hemispheric bias in gestural communication as well. Throughout evolution, there would have been selective pressure for bilateral symmetry in the limbs and organs associated with the impact of the spatial environment, because any systematic asymmetries might cause us to move in circles rather than in straight lines or to miss predators or prey lurking on one or other side. But there need be no such pressure toward symmetry in vocal control, which is relatively impervious to spatial constraints, and asymmetry may well have been selected to overcome disadvantages associated with duplication. Once gesture became associated with vocalization, it too may have become lateralized, and since gestural language is programmed internally, and is therefore independent of environmental constraints, it too may have benefited from lateralization.

My guess is that our own species, *Homo sapiens*, discovered that

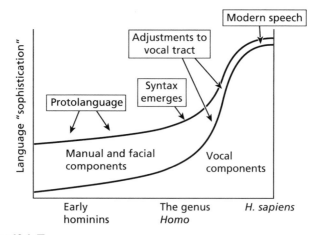

■ **Figure 10.1.** ■

A schematic representation of the development of language, and the hypothesized contributions of gestural and vocal components, in the course of hominin evolution.

language could be conveyed more or less autonomously by speech alone. It is true that we embellish our conversations with gestures, or resort to gesture alone when silence is forced upon us, whether through deafness, monastic edict, ignorance of a foreign tongue, or the proximity of predator or prey. Nevertheless, we can convey most messages by voice alone. The adaptations necessary to do this may have been in place well before our forebears discovered that it was possible, just as the adaptations necessary for writing were in place well before the early scribes scribbled or scratched their messages. The invention of autonomous speech may have been as recent as 50,000 years ago (see figure 10.1).

Molecular evidence also suggests a decisive migration of our species from Africa around 50,000 years ago that led eventually to the extinction of all previous migrants and their descendents, including not only the Neanderthals in Europe and *Homo erectus* in Asia, but also *Homo sapiens*. These newcomers may have been responsible for the explosion of art and technology that occurred in Europe starting around 40,000 years ago. The advance of technology may have been a consequence of the invention of autonomous speech, which freed the hands from involvement in communication and allowed people to speak while they engaged in manual activities. This, in turn, may have given a boost to pedagogy. The later inventions of writing, mathematics, and computer technology would have had impacts of similar magnitude. As a result of these progressive developments, culture has replaced biology as the main source of human accomplishment and variation.

So there it is—my story of how language evolved. Perhaps it is for the birds, after all. But I hope you will think it's more than mere handwaving.

■ REFERENCES ■

Abry, C., L.-J. Boë, R. Laboissière, and J.-L. Schwartz. 1998. A new puzzle for the evolution of speech. *Behavioral and Brain Sciences* 21: 512–13.

Acredolo, L., and S. Goodwyn. 1988. Symbolic gesturing in normal infants. *Child Development* 59: 450–66.

Adcock, G. J., E. S. Dennis, S. Easteal, G. A. Huttley, L. S. Jermiin, W. J. Peacock, and A. S. Thorne. 2001. Mitochondrial DNA sequences in ancient Australians: Implications for modern human origins. *Proceedings of the National Academy of Sciences* 98: 537–42.

Aich, H., R. Moos-Heilen, and E. Zimmerman, 1990. Vocalization of adult gelada baboons (*Theropithecus gelada*): Acoustic structure and behavioral context. *Folia Primatologica* 55: 109–32.

Aiello, L. C., and M. Collard. 2001. Our newest oldest ancestor? *Nature* 410: 526–27.

Ambrose, S. H. 2001. Paleolithic technology and human evolution. *Science* 291: 1748–53.

Andrews, G., P. T. Quinn, and W. A. Sorby. 1972. Stuttering: An investigation into cerebral dominance for speech. *Journal of Neurology, Neurosurgery and Psychiatry* 35: 414–18.

Andrews, P. J. 1989. Palaeoecology of Laetoli. *Journal of Human Evolution* 18: 173–81.

Annett, M. 1972. The distribution of manual asymmetry. *British Journal of Psychology* 63: 343–58.

————. 1993. The right shift theory of a genetic balanced polymorphism for cerebral dominance and cognitive processing. *Cahiers de Psychologie Cognitive* 14: 427–80.

Appenzeller, T. 1998. Art: Evolution or revolution? *Science* 282: 1451.

Arbib, M. A., and G. Rizzolatti. 1997. Neural expectations: A possible evolutionary path from manual skills to language. *Communication and Cognition* 29: 393–424.

Arcadi, A. C. 1996. Phrase structure of wild chimpanzee pant hoots: Patterns of production and interpopulation variability. *American Journal of Primatology* 39: 159–78.

————. 2000. Vocal responsiveness in male wild chimpanzees: Implications for the evolution of language. *Journal of Human Evolution* 39: 205–23.

Arcadi, A. C., D. Robert, and C. Boesch. 1998. Buttress drumming by wild chimpanzees: Temporal patterning, phrase integration into loud calls, and preliminary evidence for individual differences. *Primates* 39: 505–18.

Armstrong, D. F. 1999. *Original signs: Gesture, sign, and the source of language.* Washington, D.C.: Gallaudet University Press.

Armstrong, D. F., W. C. Stokoe, and S. E. Wilcox. 1995. *Gesture and the nature of language.* Cambridge: Cambridge University Press.

Balcombe, J. P., and M. B. Fenton. 1988. The communication role of echolocation calls in vespertilionid bats. In *Animal sonar: Processes and performances,* ed. P. E. Nachtigall and P.W.B. Moore, 625–28. New York: Plenum.

Balter, M. 1999. New light on the oldest art. *Science* 283: 920–22.

————. 2000. Paintings in Italian cave may be the oldest yet. *Science* 290: 419–21.

————. 2001a. Scientists spar over claims of earliest human ancestor. *Science* 291: 1460–61.

————. 2001b. In search of the first Europeans. *Science* 291: 1722–25.

Barakat, R. A. 1987. Cistercian sign languages. In *Monastic sign languages,* ed. J. Uniker-Sebeok and T. A. Sebeok, 69–322. Berlin: Mouton de Gruyter.

Barnett, K. J., and M. C. Corballis. 2002. Ambidexterity and magical ideation. *Laterality* 7: 75–84.

Baron-Cohen, S. 1999. The evolution of a theory of mind. In *The descent of mind,* ed. M. C. Corballis, M. C. and S.E.G. Lea, 261–77. Oxford: Oxford University Press.

Bauer, R. H. 1993. Lateralization of neural control for vocalization by the frog (*Rana pipiens*). *Psychobiology* 21: 243–48.

Bengston, J. 1992. Eve's dictionary. In *Nostratic, Dene-Caucasian, Austric, and Amerind*, ed. V. Shevoroshkin, 474–79. Bochum, Germany: Brockmeyer.

Bickerton, D. 1984. The language bioprogram hypothesis. *Behavioral and Brain Sciences* 7: 173–222.

———. 1995. *Language and human behavior.* Seattle: University of Washington Press.

Biederman, I. 1987. Recognition-by-components: A theory of human image understanding. *Psychological Review* 94: 115–47.

Bierce, A. 1997. *The devil's dictionary.* 6th ed. Oxford: Oxford University Press.

Bingham, P. M. 1999. Human uniqueness: A general theory. *Quarterly Review of Biology* 74: 133–69.

Blackmore, S. 1999. *The meme machine.* Oxford: Oxford University Press.

Blakemore, C. 1991. Computational principles of the cerebral cortex. *Psychologist* 14: 73.

Blakemore, S.-J., D. M. Wolpert, and C. D. Frith. 1998. Central cancellation of self-produced tickle sensation. *Nature Neuroscience* 1: 635–40.

Boesch-Achermann, H., and C. Boesch. 1994. Hominization in the rainforest: The chimpanzee's piece of the puzzle. *Evolutionary Anthropology* 3: 9–16.

Bogen, J. E. 1993. The callosal syndromes. In *Clinical neuropsychology*, 3d ed., ed. K. M. Heilman and E. Valenstein, 337–407. New York: Oxford University Press.

Bradshaw, J. L. 1997. *Human evolution: A neuropsychological perspective.* Hove, U.K.: Psychology.

Bradshaw, J. L., and L. J. Rogers. 1993. *The evolution of lateral asymmetries, language, tool use, and intellect.* Sydney: Academic.

Brain, C. K. 1993. The occurrence of burnt bones at Swartkrans and their implications for the control of fire by early humans. In *Swartkrans: A cave's chronicle of early man*, ed. C. K. Brain, 229–42. Transvaal Museum monograph no. 8. Pretoria: Transvaal Museum.

Broadhurst, C. L., S. C. Cunnane, and M. A. Crawford. 1998. Rift Valley lake-fish and shellfish provided brain-specific nutrition for early *Homo*. *British Journal of Nutrition* 79: 3–21.

Browman, C., and L. Goldstein. 1991. Gestural structures: Distinctiveness, phonological processes, and historical change. In *Modularity and the motor theory of speech perception*, ed. I. G. Mattingly and M. Studdert-Kennedy, 313–38. Hillsdale, N.J.: Erlbaum.

Brown, F., J. Harris, R. Leakey, and A. Walker. 1985. Early *Homo erectus* skeleton from west Lake Turkana, Kenya. *Nature* 316: 788–92.

Brueckner, M., P. D'Eustachio, and A. L. Horwich. 1989. Linkage mapping of a mouse gene, iv, that controls left-right asymmetry of the heart and viscera. *Proceedings of the National Academy of Sciences* 86: 5035–38.

Brugger, P., A. Gamma, R. Muri, M. Schaffer, and K. T. Taylor. 1993. Functional hemispheric asymmetry and belief in ESP: Towards a 'neuropsychology of belief.' *Perceptual and Motor Skills* 77: 1299–1308.

Brunet, M., A. Beauvilain, Y. Coppens, E. Heintz, A.H.E. Moutaye, and D. Pilbeam. 1995. The first australopithecine 2,500 kilometers west of the Rift Valley (Chad). *Nature* 378: 273–75.

Buettner-Janusch, J. 1966. *Origins of man.* New York: Wiley.

Burling, R. 1999. Motivation, conventionalization, and arbitrariness in the origin of language. In *The origins of language: What nonhuman primates can tell us,* ed. B. J. King, 307–50. Santa Fe, N.M.: School of American Research Press.

Butler, D. 2001. The battle of Tugen Hills. *Nature* 410: 508–9.

Butterworth, B., and U. Hadar. 1989. Gesture, speech, and computational stages: A reply to McNeill. *Psychological Review* 96: 168–74.

Buxhoeveden, D., and M. Casanova. 2000. Comparative lateralisation patterns in the language area of human, chimpanzee, and rhesus monkey brains. *Laterality* 4: 315–30.

Byrne, R. W. 1995. *The thinking ape.* Oxford: Oxford University Press.

———. 1996. The misunderstood ape: Cognitive skills of the gorilla. In *Reaching into thought,* ed. A. E. Russon, K. A. Bard, and S. T. Parker, 111–30. Cambridge: Cambridge University Press.

———. 2000. Evolution of primate cognition. *Cognitive Science* 24: 543–70.

Byrne, R. W., and A. E. Russon. 1998. Learning by imitation: A hierarchical approach. *Behavioral and Brain Sciences* 21: 667–721.

Calvin, W. H. 1983. *The throwing madonna: Essays on the brain.* New York: McGraw-Hill.

Calvin, W. H., and D. Bickerton. 2000. *Lingua ex machina: Reconciling Darwin and Chomsky with the human brain.* Cambridge, Mass.: MIT Press.

Cann, R. L. 1993. Human dispersal and divergence. *Trends in Ecology and Evolution* 8: 27–31.

———. 2001. Genetic clues to dispersal in human populations: retracing the past from the present. *Science* 291: 1742–748.

Carstairs-McCarthy, A. 1999. *The origins of complex language.* Oxford: Oxford University Press.

Cavalli-Sforza, L. L. 2000. *Genes, peoples, and languages.* New York: North Point.

Chagnon, N. A. 1968. *Yanomamö: The fierce people*. New York: Holt, Rinehart, and Winston.

———. 1988. Life histories, blood revenge, and warfare in a tribal population. *Science* 239: 985–92.

Cheney, D. L., and R. S. Seyfarth. 1990. *How monkeys see the world*. Chicago: University of Chicago Press.

Chomsky, N. 1966. *Cartesian linguistics: A chapter in the history of rational thought*. New York: Harper and Row.

———. 1986. *Knowledge of language: Its nature, origin and use*. New York: Praeger.

———. 1995. *The minimalist program*. Cambridge, Mass.: MIT Press.

———. 2000. *New horizons in the study of language and mind*. Cambridge: Cambridge University Press.

Clahsen, H. 1999. Lexical entries and rules of language. *Brain and Behavioral Sciences* 22: 991–1060.

Clark, C., and R. W. Wrangham. 1993. Acoustic analysis of wild chimpanzee hoots: Do Kibale Forest chimpanzees have an acoustically distinct food arrival pant hoot? *American Journal of Primatology* 31: 99–110.

Clayton, N. S., and A. Dickinson. 1998. Episodic-like memory during cache recovery by scrub jays. *Nature* 395: 273–78.

Collins, R. L. 1970. The sound of one paw clapping: An inquiry into the origins of left handedness. In *Contributions to behavior-genetic analysis—The mouse as a prototype*, ed. G. Lindzey and D. D. Thiessen, 115–36. New York: Meredith.

Condillac, E. Bonnot de. 1971. *An essay on the origin of human knowledge; being a supplement to Mr. Locke's Essay on the human understanding*. A facsim. reproduction of the translation of Thomas Nugent, with an introd. by Robert G. Weyant. Gainesville, Fla: Scholars' Facsimiles and Reprints.

Conrad, R. 1979. *The deaf schoolchild: Language and cognitive function*. London: Harper and Row.

Coppens, Y. 1994. East Side Story: The origins of humankind. *Scientific American* 270 (5): 88–95.

Corballis, M. C. 1991. *The lopsided ape*. New York: Oxford University Press.

———. 1992. On the evolution of language and generativity. *Cognition* 44: 197–226.

———. 1997. The genetics and evolution of handedness. *Psychological Review* 104: 714–27.

———. 1998. Cerebral asymmetry: Motoring on. *Trends in Cognitive Science* 2: 152–57.

———. 1999a. Are we in our right minds? In *Mind myths*, ed. S. Della Sala, 25–41. New York: Wiley.

———. 1999b. The gestural origins of language. *American Scientist* 87 (March–April): 38–145.

———. 2001. Is the handedness gene on the X chromosome? *Psychological Review* 108: 805–10.

Corballis, M. C., and M. J. Morgan. 1978. On the biological basis of human laterality: I. Evidence for a maturational left-right gradient. *The Behavioral and Brain Sciences* 2: 261–68.

Coren, S., and D. F. Halpern. 1991. Left-handedness: A marker for decreased survival fitness. *Psychological Bulletin* 109: 90–106.

Coren, S., and C. Porac. 1977. Fifty centuries of right-handedness: The historical record. *Science* 198: 631–32.

Corina, D. P., D. Bavelier, and H. J. Neville. 1998. Response from Corina, Neville and Bavelier. *Trends in Cognitive Sciences* 2: 468–70.

Coulmas, F. 1989. *The writing systems of the world.* Oxford: Blackwell.

Cox, J. R., and R. A. Griggs. 1989. The effects of experience on performance in Wason's selection tasks. *Memory and Cognition* 10: 496–502.

Crain, S., and M. Nakayama. 1986. Structure dependence in children's language. *Language* 62: 522–43.

Critchley, M. 1939. *The language of gesture.* London: Arnold.

———. 1975. *Silent language.* London: Butterworths.

Crow, T. J. 1997. Schizophrenia as failure of hemispheric dominance for language. *Trends in Neurosciences* 20: 339–43.

———. 1998. Sexual selection, timing and the descent of man: A theory of the genetic origins of language. *Current Psychology of Cognition* 17: 1237–77.

Crow, T. J., L. R. Crow, D. J. Done, and S. Leask. 1998. Relative hand skill predicts academic ability: global deficits at the point of hemispheric indecision. *Neuropsychologia* 36: 1275–82.

Crystal, D. 1997. *The Cambridge encyclopedia of language.* 2d ed. Cambridge: Cambridge University Press.

Dart, R. A. 1925. *Australopithecus africanus*: The man-ape of South Africa. *Nature* 115: 195–99.

Darwin, C. 1859. *The origin of species.* London: John Murray.

———. 1896. *The descent of man and selection in relation to sex.* London: William Clowes. Orig. pub. in 1871.

———. 1965. *The expression of the emotions in man and animals.* Chicago: University of Chicago Press. Orig. pub. London: John Murray, 1904.

Davidson, I., and W. Noble. 1992. Why the first colonisation of the Australian region is the earliest evidence of modern human behaviour. *Perspectives in Human Biology* 2: 135–42.

———. 1993. Tools and language in human evolution. In *Tools, language, and cognition in human evolution*, ed. K. R. Gibson and T. Ingold, 363–88. Cambridge: Cambridge University Press.

Dawkins, R. 1976. *The selfish gene.* Oxford: Oxford University Press.

Dawkins, R., and J. R. Krebs. 1978. Animal signals: information or manipulation? In *Behavioral ecology: An evolutionary approach*, ed. J. R. Krebs and N. B. Davies, 283–309. Sunderland, Mass.: Sinauer.

Dawson, J. L. 1972. Temne-Arunta hand-eye dominance and cognitive style. *International Journal of Psychology* 7: 219–33.

Deacon, T. 1997. *The symbolic species.* New York: Norton.

Dennett, D. 1992. *The role of language in intelligence.* Publication CCS-92-3, Center for Cognitive Studies, Tufts University, Boston, Mass.

Descartes, R. 1985. *The philosophical writings of Descartes.* Ed. and trans. J. Cottingham, R. Stoothoff, and D. Murdock. Cambridge: Cambridge University Press. Orig. pub. 1647.

Deuchar, M. 1996. Spoken language and sign language. In *Handbook of symbolic evolution*, ed. A. Lock and C. R. Peters, 553–70. Oxford: Clarendon Press.

De Villiers, J. G., and P. A. de Villiers. 1999. Linguistic determinism and the understanding of false beliefs. In *Children's reasoning and the understanding of false beliefs*, ed. P. Mitchell and K. Riggs, 198–228. Hove, U.K.: Psychology Press.

De Waal, F. 1982. *Chimpanzee politics.* New York: Harper and Row.

Diamond, A. S. 1959. *The history and origin of language.* London: Methuen.

Diamond, J. 1992. *The third chimpanzee.* New York: HarperCollins.

———. 1997. *Guns, germs, and steel: The fates of human societies.* New York: Norton.

———. 2001. Unwritten knowledge. *Nature* 410: 521.

Dominguez-Rodrigo, M., J. Serrallonga, J. Juan-Tresserras, L. Alcala, and L. Luque, 2001. Woodworking activities by early humans: A plant residue analysis on Acheulian stone tools from Peninj (Tanzania). *Journal of Human Evolution* 40: 289–99.

Donald, M. 1991. *Origins of the modern mind: Three stages in the evolution of culture and cognition.* Cambridge, Mass.: Harvard University Press.

———. 1999. Preconditions for the evolution of protolanguages. In *The descent of mind*, ed. M. C. Corballis and S.E.G. Lea, 138–54. Oxford: Oxford University Press.

Dunbar, R.I.M. 1992. Social behaviour and evolutionary theory. In *The Cam-*

bridge encyclopedia of human evolution, ed. S. Jones, R. Martin, and D. Pilbeam, 145–47. Cambridge: Cambridge University Press.

———. 1993. Coevolution of neocortical size, group size and language in humans. *Behavioral and Brain Sciences* 16: 681–735.

Dunbar, R.I.M., and L. C. Aiello. 1993. Neocortex size, group size, and the evolution of language. *Current Anthropology* 34: 184–93.

Dunbar, R.I.M., and P. Dunbar. 1975. Social dynamics of gelada baboons. In *Contributions to primatology*, vol. 6, ed. F. S. Szalay. New York: Karger.

Eccles, J. C. 1965. *The brain and the unity of conscious experience.* Cambridge: Cambridge University Press.

———. 1981. Mental dualism and commissurotomy. *Behavioral and Brain Sciences* 4: 105.

Ehert, G. 1987. Left hemisphere advantage in the mouse brain for recognizing ultrasonic communication calls. *Nature* 325: 249–51.

Elman, J., E. Bates, and E. Newport. 1996. *Rethinking innateness: A connectionist perspective on development.* Cambridge: MIT Press.

Fagan, B. 1989. *People of the earth: An introduction to world prehistory*, 6th ed. Glenview, Ill.: Scott, Foresman.

Falk, D. 1975. Comparative anatomy of the larynx in man and chimpanzee: Implications for language in Neanderthal. *American Journal of Physical Anthropology* 43: 123–32.

———. 1983. The Taung endocast: A reply to Holloway. *American Journal of Physical Anthropology* 60: 17–45.

Fernandes, M.E.B. 1991. Tool use and predation of oysters by the tufted capuchin in brackish water mangrove swamp. *Primates* 32: 529–31.

Feyereisen, P. 1997. The competition between gesture and speech production in dual-task paradigms. *Journal of Memory and Language* 36: 13–33.

Fifer, F. C. 1987. The adoption of bipedalism by the hominids: A new hypothesis. *Human Evolution* 2: 135–47.

Fitch, R. H., C. P. Brown, K. O'Connor, and P. Tallal. 1993. Function lateralization for auditory temporal processing in male and female rats. *Behavioral Neuroscience* 107: 844–50.

Fitch, W. T. 2000. The evolution of speech: A comparative review. *Trends in Cognitive Sciences* 4: 258–66.

Fodor, J. 2000. *The mind doesn't work that way.* Cambridge, Mass.: MIT Press.

Frishberg, N. 1975. Arbitrariness and iconicity in American Sign Language. *Language* 51: 696–719.

Funnell, M. G., P. M. Corballis, and M. S. Gazzaniga. 1999. A deficit in perceptual matching in the left hemisphere of a callosotomy patient. *Neuropsychologia* 37: 1143–54.

Gabunia, L., A. Vekua, D. Lordkipanidze, C. C. Swisher, III, R. Ferring, A. Justus, M. Nioradze, M. Tvalchrelidze, S. C. Anton, G. Bosinksi, O. Joris, M.-A. de Lumley, G. Majsuradze, and A. Mouskhelishvili. 2000. Earliest Pleistocene hominid cranial remains from Dmanisi, Republic of Georgia: Taxonomy, geological setting, and age. *Science* 288: 1019–25.

Gallese, V., and A. Goldman. 1998. Mirror neurons and the simulation theory of mind reading. *Trends in Cognitive Science* 2: 493–501.

Gannon, P. J., R. L. Holloway, D. C. Broadfield, and A. R. Braun, 1998. Asymmetry of chimpanzee planum temporale: Human-like brain pattern of Wernicke's area homolog. *Science* 279: 220–21.

Gardner, M. 1960. *The annotated Alice.* New York: Charles N. Potter.

———. 1967. *The ambidextrous universe.* London: Penguin.

Gardner, R. A., and B. T. Gardner. 1969. Teaching sign language to a chimpanzee. *Science* 165: 664–72.

Gaur, A. 1987. *A history of writing.* London: British Library.

Gazzaniga, M. S. 1987. Perceptual and attentional processes following callosal section in humans. *Neuropsychologia* 25: 119–33.

———. 2000. Cerebral specialization and interhemispheric communication: Does the corpus callosum enable the human condition? *Brain* 123: 1293–26.

Gelb, I. J. 1974. *A study of writing.* Chicago: University of Chicago Press.

Geschwind, N., and W. Levitsky. 1968. Human brain: Left-right asymmetries in temporal speech region. *Science* 161: 186–87.

Gibson, K. R., and S. Jessee, 1999. Language evolution and expansions of multiple neurological processing areas. In *The origins of language: What nonhuman primates can tell us,* ed. B. J. King, 189–227. Santa Fe, N.M.: School of American Research Press.

Givón, T. 1995. *Functionalism and grammar.* Philadelphia: John Benjamins.

Goldin-Meadow, S., and D. McNeill. 1999. The role of gesture and mimetic representation in making language the province of speech. In *The descent of mind,* ed. M. C. Corballis and S.E.G. Lea, 155–72. Oxford: Oxford University Press.

Goldin-Meadow, S., D. McNeill, and J. Singleton. 1996. Silence is liberating: Removing the handcuffs on grammatical expression and speech. *Psychological Review* 103: 34–55.

Goldin-Meadow, S., and C. Mylander. 1998. Spontaneous sign systems created by deaf children in two cultures. *Nature* 391: 279–81.

Goller, F., and R. A. Suthers. 1995. Implications for lateralization of bird song from unilateral gating of bilateral motor patterns. *Nature* 373: 63–65.

Goodall, J. 1986. *The chimpanzees of Gombe: Patterns of behavior.* Cambridge, Mass.: Harvard University Press.

Goren-Inbar, N., C. S. Feibel, K. L. Verosub, Y. Melamed, M. E. Kislev, E. Tchernov, and I. Saragusti. 2000. Pleistocene milestones on the out-of-Africa corridor at Gesher Benot Ya'aqov, Israel. *Science* 289: 944–47.

Gould, S. J., and R. C. Lewontin. 1979. The spandrels of San Marco and the Panglossian programme: A critique of the adaptationist programme. *Proceedings of the Royal Society of London* 205: 281–88.

Gould, S. J., and E. S. Vrba. 1982. Exaptation—a missing term in the science of form. *Paleobiology* 8: 4–15.

Gowlett, J.A.J. 1984. *Ascent to civilization.* London: Collins.

Greenfield, P. M. 1991. Language, tools, and the brain: The ontogeny and phylogeny of hierarchically organized sequential behavior. *Behavioral and Brain Sciences* 14: 531–95.

Greenfield, P. M., A. E. Maynard, C. Boehm, and E. Y. Schmidtling. 2000. Cultural apprenticeship and cultural change. In *Biology, brains, and behavior*, ed. S. T. Parker, J. Langer and M. L. McKinney, 237–77. Santa Fe, N.M.: School of American Research Press.

Griffin, D. R. 1976. *The question of animal awareness.* New York: Rockefeller University Press.

Grossman, M. 1980. A central processor for hierarchically structured material: Evidence from Broca's aphasia. *Neuropsychologia* 18: 299–308.

Groves, C. P. 1989. *A theory of human and primate evolution.* Oxford: Clarendon.

Haimoff, E. H. 1986. Convergence in the duetting of monogamous old world primates. *Journal of Human Evolution* 15: 51–59.

Hamilton, W. D. 1964. The genetic evolution of social behavior, I. *Journal of Theoretical Biology* 7: 1–16.

Hammer, M. F. 1995. A recent common ancestry for human Y chromosomes. *Nature* 378: 376–78.

Hardy, A. 1960. Was man more aquatic in the past? *New Scientist* 7: 642–45.

Hare, B., J. Call, B. Agnetta, and M. Tomasello. 2000. Chimpanzees know what conspecifics do and do not see. *Animal Behaviour* 59: 771–85.

Hare, B., and M. Tomasello. 1999. Domestic dogs (*Canis familiaris*) use human and conspecific cues to locate hidden food. *Journal of Comparative Psychology* 113: 173–77.

Harris, L. J. 1993. Do left-handers die sooner than right-handers? Commentary on Coren and Halpern's (1991) "Left-handedness: A marker for decreased survival fitness." *Psychological Bulletin* 114: 203–34.

Hauser, M. D. 1996. *The evolution of communication.* Cambridge, Mass.: MIT Press.

Hauser, M. D., and K. Anderson. 1994. Functional lateralization for auditory temporal processing in adult, but not infant, rhesus monkeys: Field experiments. *Proceedings of the National Academy of Sciences* (U.S.) 91: 3946–48.

Hauser, M. D., P. Teixidor, L. Field, and R. Flaherty. 1993. Food-elicited calls in chimpanzees: Effects of food quantity and divisibility. *Animal Behaviour* 45: 817–19.

Hauser, M. D., and R. W. Wrangham. 1987. Manipulation of food calls in captive chimpanzees: A preliminary report. *Folio Primatologica* 48: 207–10.

Hayes, C. 1952. *The ape in our house.* London: Gollancz.

Hedges, S. B. 2000. A start for population genomics. *Nature* 408: 652–53.

Heffner, H. E., and R. S. Heffner. 1984. Temporal lobe lesions and perception of species-specific vocalizations by Japanese macaques. *Science* 226: 75–76.

Heilman, K. M., and L.J.G. Rothi. 1985. Apraxia. In *Clinical neuropsychology,* 2d ed., K. M. Heilman and E. Valenstein, 131–50. Oxford: Oxford University Press.

Hewes, G. W. 1973. Primate communication and the gestural origins of language. *Current Anthropology* 14: 5–24.

———. 1981. Pointing and language. In *The cognitive representation of speech,* ed. T. Myers, J. Laver, and J. Anderson, 263–69. Amsterdam: North-Holland.

———. 1996. A history of the study of language origins and the gestural primacy hypothesis. In *Handbook of human symbolic evolution,* ed. A. Lock and C. R. Peters, 571–95. Oxford: Oxford University Press.

Hickok, G., U. Bellugi, E. S. and Klima. 1997. The basis of the neural organization for language: Evidence from sign language aphasia. *Reviews in the Neurosciences* 8: 205–22.

———. 1998. What's right about the neural organization of sign language? A perspective on recent neuroimaging results. *Trends in Cognitive Sciences* 2: 465–68.

Highfield, R., and P. Carter. 1993. *The private lives of Albert Einstein.* New York: St. Martin's.

Hockett, C. F. 1960a. The origin of speech. *Scientific American* 203: 88–96.

———. 1960b. Logical considerations in the study of animal communication. In *Animal sounds and communication,* Symposium Series no. 7, ed. W. E. Landon and W. N. Tavolga, 392–430. Washington, D.C.: American Institute of Biological Sciences.

Holloway, R. L. 1969. Culture: A human domain. *Current Anthropology* 4: 135–68.

———. 1983. Human paleontological evidence relevant to language behavior. *Human Neurobiology* 2: 105–14.

———. 1985. The past, present, and future significance of the lunate sulcus in early hominid evolution. In *Hominid evolution: Past, present, and future*, ed. P. V. Tobias, 47–62. New York: Allen R. Liss.

Hopkins, W. D. 1996. Chimpanzee handedness revisited: 55 years since Finch (1941). *Psychonomic Bulletin and Review* 3: 449–57.

Hopkins, W. D., and D. A. Leavens. 1998. Hand use and gestural communication in chimpanzees (Pan troglodytes). *Journal of Comparative Psychology* 112: 95–99.

Hudson, R. A. 1980. *Sociolinguistics*. Cambridge: Cambridge University Press.

Hunt, G. R. 2000. Human-like, population-level specialization in the manufacture of pandanus tools by New Caledonian crows *Corvus moneduloides*. *Proceedings of the Royal Society of London B* 267: 403–13.

Imanishi, K. 1957. Identification: A process of enculturation in the subhuman society of *Macaca fuscata*. *Primates* 1: 1–29.

Ingman, M., H. Kaessmann, S. Pääbo, and U. Gyllensten. 2000. Mitochondrial genome variation and the origin of modern humans. *Nature* 408: 708–13.

Isaac, B. 1987. Throwing and human evolution. *Archeological Review* 5: 3–17.

———. 1992. Throwing. In *The Cambridge encyclopedia of human evolution*, ed. S. Jones, R. Martin, and D. Pilbeam, 358. Cambridge: Cambridge University Press.

Iverson, J. M., O. Capirci, and M. C. Caselli. 1994. From communication to language in two modalities. *Cognitive Development* 9: 23–43.

Jakobson, R. 1940. Kindersprache, Aphasie und allgemeine Lautgesetze. In *Selected writings*. The Hague: Mouton. Trans. A. R. Keller, in *Child language, aphasia and phonological universals*. The Hague: Mouton, 1968.

Jalles-Filho, E., R.G.T. Da Cunha, and R. A. Salm. 2001. Transport of tools and mental representation: Is capuchin monkey tool behavior a useful model of Plio-Pleistocene hominid technology? *Journal of Human Evolution* 40: 365–77.

Jäncke, L., and H. Steinmetz. 1993. Auditory lateralization and planum temporale asymmetry. *Neuroreport* 5: 169–72.

Janik, V. M. 2000. Whistle matching in wild bottlenose dolphins (*Tursiops truncatus*). *Science* 289: 1355–57.

Jaynes, J. 1976. *The origins of consciousness in the breakdown of the bicameral mind*. Boston: Houghton Mifflin.

Johanson, D. C., and M. A. Edey. 1981. *Lucy, the beginnings of humankind*. New York: Simon and Schuster.

Johnson, F. 2000. I'm a soul man. *Spectator* 284 (8958l): 10–11.

Johnson, W. E., ed. 1955. *Stuttering in children and adults*. Minneapolis: University of Minnesota Press.

Jones, G. V., and M. Martin. 2000. A note on Corballis (1997) and the genetics and evolution of handedness: Developing a unified distributional model from the sex-chromosomes gene hypothesis. *Psychological Review* 107: 213–18.

Jones, S. 2000. *Almost like a whale*. London: Anchor Books.

Jones, S., R. Martin, and D. Pilbeam, eds. 1992. *The Cambridge encyclopedia of human evolution*. Cambridge: Cambridge University Press.

Jürgens, U., A. Kirzinger, and D. von Cramon. 1982. The effect of deep reaching lesions in the cortical face area on phonation: A combined case report and experimental monkey study. *Cortex* 18: 125–40.

Karnath, H.-O., S. Ferber, and M. Himmelbach. 2001. Spatial awareness is a function of the temporal lobe not the posterior parietal lobe. *Nature* 411: 950–53.

Kay, R. F., M. Cartmill, and M. Barlow. 1998. The hypoglossal canal and the origin of human vocal behavior. *Proceedings of the National Academy of Sciences* (U.S.) 95: 5417–19.

Kay, R. F., J. G. Fleagle, and E. L. Symonds. 1981. A revision of the Oligene apes from the Fayum Province, Egypt. *American Journal of Physical Anthropology* 55: 293–322.

Ke, Y., B. Su, X. Song, D. Lu, L. Chen, H. Li, C. Qi, S. Marzuki, R. Deka, P. Underhill, C. Xiao, M. Shriver, J. Lell, D. Wallace, R. S. Wells, M. Seielstad, P. Oefner, D. Zhu, J. Jin, W. Huang, R. Chakraborty, Z. Chen, and L. Jin. 2001. African origin of modern humans in East Asia: A tale of 12,000 Y chromosomes. *Science* 292: 1151–53.

Kendon, A. 1988. *Sign languages of aboriginal Australia*. Melbourne: Cambridge University Press.

Kimura, D. 1973a. Manual activity while speaking: I. Right-handers. *Neuropsychologia* 11: 45–50.

———. 1973b. Manual activity while speaking: II. Left-handers. *Neuropsychologia* 11: 51–55.

———. 1981. Neural mechanisms of manual signing. *Sign Language Studies* 33: 291–312.

————. 1999. *Sex and cognition.* Cambridge, Mass.: MIT Press.

Kinsbourne, M. 1978. Evolution of language in relation to lateral action. In *Asymmetrical function of the brain,* ed. M. Kinsbourne, 553–65. Cambridge: Cambridge University Press.

Knecht, S., B. Dräger, M. Deppe, L. Bobe, H. Lohmann, A. Flöel, E.-B. Ringelstein, and H. Henningsen. 2000. Handedness and hemispheric language dominance in healthy humans. *Brain* 123: 2512–18.

Knight, C. 1998. Ritual/speech coevolution: A solution to the problem of deception. In *Approaches to the evolution of language,* ed. J. R. Hurford, M. Studdert-Kennedy, and C. Knight, 68–91. Cambridge: Cambridge University Press.

Kobayashi, H. and S. Kohshima. 2001. Unique morphology of the human eye and its adaptive meaning: Comparative studies on external morphology of the human eye. *Journal of Human Evolution* 40: 419–35.

Köhler, W. 1925. *The mentality of apes.* New York: Routledge and Kegan Paul.

Langdon, J. H. 1993. Umbrella hypotheses and parsimony in human evolution: A critique of the aquatic ape hypothesis. *Journal of Human Evolution* 33: 479–94.

Larson, D. B., J. P. Swyers, and McCullough, M. E., eds. 1998. *Scientific research on spirituality and health: A consensus report.* Rockville, Md.: National Institute for Healthcare Research.

Leakey, M. D., C. S. Feibel, I. McDougall, and A. Walker. 1995. New four-million-year-old hominid species from Kanapoi and Allia Bay, Kenya. *Nature* 376: 565–71.

Leakey, M. D., and J. M. Harris, eds. 1987. *Laetoli: A Pliocene site in northern Tanzania.* Oxford: Clarendon Press.

Leakey, M. G., F. Spoor, F. Brown, P. N. Gathogo, C. Kiarie, L. N. Leakey, and I. McDougall. 2001. New hominin genus from eastern Africa shows diverse middle Pliocene lineages. *Nature* 410: 433–40.

Leavens, D. A., and W. D. Hopkins. 1998. Intentional communication by chimpanzees: A cross-sectional study of the use of referential gestures. *Developmental Psychology* 34: 813–22.

Lee, R. B. 1979. *The !Kung San: Men, women, and work in a foraging society.* Cambridge: Cambridge University Press.

Leonhard, D., and P. Brugger. 1998. Creative, paranormal and delusional thought: A consequence of right hemispheric semantic activation? *Neuropsychiatry Neuropsychology and Behavioral Neurology* 11: 177–83.

Levelt, W.J.M. 2000. Psychology of language. In *International Handbook of Psychology*, ed. K. Pawlik and M. R. Rosenzweig. London: Sage.

Lewin, R. 1988. Linguists search for the mother tongue. *Science* 242: 1128–29.

Liberman, A. M., F. S. Cooper, D. P. Shankweiler, and M. Studdert-Kennedy, 1967. Perception of the speech code. *Psychological Review* 74: 431–61.

Liberman, A. M., and I. G. Mattingly. 985. The motor theory of speech perception revisited. *Cognition* 21: 1–36.

Liberman, A. M., and D. H. Whalen. 2000. On the relation of speech to language. *Trends in Cognitive Sciences* 4: 187–96.

Lieberman, D. E. 1998. Sphenoid shortening and the evolution of modern human cranial shape. *Nature* 393: 158–62.

Lieberman, P. 1982. Can chimpanzees swallow or talk? A reply to Falk. *American Anthropologist* 84: 148–52.

———. 1998. *Eve spoke: Human language and human evolution.* New York: Norton.

———. 2000. *Human language and our reptilian brain.* Cambridge, Mass.: Harvard University Press.

Lieberman, P., E. S. Crelin, and D. H. Klatt. 1972. Phonetic ability and related anatomy of the new-born, adult human, Neanderthal man, and the chimpanzee. *American Anthropologist* 74: 287–307.

Lock, A. J. 1980. *The guided reinvention of language.* London: Academic Press.

———. 1999. On the recent origin of symbolically-mediated language and its implications for psychological science. In *The descent of mind,* ed. M. C. Corballis, and S.E.G. Lea, 324–55. Oxford: Oxford University Press.

MacLarnon, A., and G. Hewitt. 1999. The evolution of human speech: The role of enhanced breathing control. *American Journal of Physical Anthropology* 109: 341–63.

MacNeilage, P. F. 1998. The frame/content theory of the evolution of speech production. *Behavioral and Brain Sciences* 21: 499–56.

MacNeilage, P. F., M. G. Studdert-Kennedy, and B. Lindblom. 1987. Primate handedness reconsidered. *Behavioral and Brain Sciences* 10: 247–303.

Maess, B., S. Koelsch, T. C. Gunter, and A. D. Friederici. 2001. Musical syntax is processed in Broca's area: An MEG study. *Nature Neuroscience* 4: 540–45.

Mallery, G. 1880. *A collection of gesture-signs and signals of North American*

Indians with some comparisons. Washington, D.C.: Government Printing Office. Repr. in Umiker-Sebeok and Sebeok, 1978.

Marquez, G. G. 1971. *One hundred years of solitude*. New York: Avon.

Marshall, A. J., R. W. Wrangham, and A. C. Arcadi. 1999. Does learning affect the structure of vocalizations in chimpanzees? *Animal Behaviour* 58: 825–30.

Martineau, L. 1973. *The rocks begin to speak*. Las Vegas: K. C. Publications.

Marzke, M. W. 1996. Evolution of the hand and bipedality. In *Handbook of symbolic evolution*, ed. A. Lock and C. R. Peters, 126–54. Oxford: Oxford University Press.

Mayberry, R. I., and E. Nicolaidis. 2000. Gesture reflects language development: Evidence from bilingual children. *Current Directions in Psychological Science* 9: 192–96.

Mayberry, R. I., and R. C. Shenker. 1997. Gesture mirrors speech motor control in stutterers. In *Speech motor production and fluency disorders*, ed. W. Hulstijn, H. Peters, and P. van Lieshout, 183–90. Amsterdam: Elsevier Science.

Maynard Smith, J., and E. Szathmáry. 1995. *The major transitions in evolution*. New York: W. H. Freeman Spektrum.

Mazzarello, P. 2001. Lombroso and Tolstoy. *Science* 409: 983.

McCarthy, M. 1957. *Memories of a Catholic girlhood*. London: Heinemann.

McClelland, J. L., and M. S. Seidenberg. 2000. Why do kids say goed and brang? Review of S. Pinker, *Words and rules. Science* 287: 47–48.

McGrew, W. C., and L. F. Marchant. 1997. On the other hand: Current issues in a meta-analysis of the behavioural laterality of hand function in nonhuman primates. *Yearbook of Physical Anthropology* 40: 201–32.

McGrew, W. C., C. E. Tutin, and P. J. Baldwin. 1979. Chimpanzees, tools, and termites: Cross-cultural comparisons of Senegal, Tanzania, and Rio Muni. *Man: The Journal of the Royal Anthropological Institute* 14: 185–214.

McGuire, P. K., D. Robertson, A. Thacker, A. S. David, N. Kitson, R.S.J. Frackowiak, and C. D. Frith. 1997. Neural correlates of thinking in sign language. *Neuroreport* 8: 695–98.

McGurk, H., and J. MacDonald. 1976. Hearing lips and seeing voices. *Nature* 264: 746–48.

McKeever, W. F. 2000. A new family handedness sample with findings consistent with X-linked transmission. *British Journal of Psychology* 91: 21–39.

McManus, I. C. 1985. Handedness, language dominance, and aphasia: A genetic model. *Psychological Medicine*, suppl. 8: 1–40.

McManus, I. C., and M. P. Bryden. 1992. The genetics of handedness, cerebral dominance, and lateralization. In *Handbook of neuropsychol-*

ogy, Vol. 6: *Developmental neuropsychology, Part 1*, ed. I. Rapin and S. J. Segalowitz, 115–144. Amsterdam: Elsevier Science.

McNeill, D. 1985. So you think gestures are nonverbal? *Psychological Review* 92: 350–71.

Mead, J., A. Bouhuys, and D. F. Proctor. 1968. Mechanisms generating subglottic pressure. *Annals of the New York Academy of Sciences* 155: 177–82.

Meier, R. P.,and E. L. Newport. 1990. Out of the hands of babes: On a possible sign language advantage in language acquisition. *Language* 66: 1–23.

Meier, R. P., and R. Willerman. 1995. Prelinguistic gesture in deaf and hearing infants. In *Language, gesture, and space*, ed. K. Emmorey and K. Reilly, 391–409. Hillsdale, N.J.: Erlbaum.

Mellars, P. 1989. Major issues in the emergence of modern humans. *Current Anthropology* 30: 349–85.

Miles, H. L. 1990. The cognitive foundations for reference in a signing orangutan. In *Language and intelligence in monkeys and apes*, ed. S. T. Parker and K. R. Gibson, 511–39. Cambridge: Cambridge University Press.

Mitani, J. C. 1996. Comparative studies of African ape vocal behavior. In *Great ape societies*, ed. W. C. McGrew, L. F. Marchant, and T. Nishida, 241–54. Cambridge: Cambridge University Press.

Mithen, S. 1996. *The prehistory of the mind.* London: Thames and Hudson.

Mithun, M. 1999. *The languages of Native North America.* Cambridge: Cambridge University Press.

Moore, C., and V. Corkum. 1998. Infant gaze-following based on eye direction. *British Journal of Developmental Psychology* 16: 495–503.

Morange-Majoux, F., A. Peze, and H. Bloch. 2000. Organisation of left and right hand movement in a prehension task: A longitudinal study from 20 to 32 weeks. *Laterality* 4: 351–62.

Morford, J. P., J. L. Singleton, and S. Goldin-Meadow. 1995. The genesis of language: How much time is needed to generate arbitrary symbols in a sign language? In *Language, gesture, and space*, ed. K. Emmorey and K. Reilly, 313–32. Hillsdale, N.J.: Erlbaum.

Morgan, E. 1990. *The scars of evolution.* New York: Oxford University Press.

———. 1997. *The aquatic ape hypothesis.* London: Souvenir.

Morwood, M. J., P. B. O'Sullivan, F. Aziz, and A. Raza. 1998. Fission-track ages of stone tools and fossils on the east Indonesian island of Flores. *Nature* 392: 173–76.

Muller, F. M. 1880. *Lectures on the science of language.* Vol. 1. London: Longmans, Green. Orig. pub. in German, 1861.

Murphy, K., D. R. Corfield, A. Guz, G. R. Fink, R.J.S. Wise, J. Harrison, and L.

Adams. 1997. Cerebral areas associated with motor control of speech in humans. *Journal of Applied Physiology* 83: 1438–47.

Neidle, C., J. Kegl, D. MacLaughlin, B. Bahan, and R. G. Lee. 2000. *The syntax of American Sign Language*. Cambridge, Mass.: MIT Press.

Neville, H. J., D. Bavelier, D. Corina, J. Rauschecker, A. Karni, A. Lalwani, A. Braun, V. Clark, P. Jezzard, and R. Turner. 1997. Cerebral organization for deaf and hearing subjects: Biological constraints and effects of experience. *Proceedings of the National Academy of Sciences* (U.S.) 95: 922–29.

Nishitani, N., and R. Hari. 2000. Dynamics of cortical representation for action. *Proceedings of the National Academy of Sciences*, (U.S.) 97: 913–18.

Noble, W., and I. Davidson. 1996. *Human evolution, language and mind*. Cambridge: Cambridge University Press.

Nottebohm, F. 1977. Asymmetries for neural control of vocalization in the canary. In *Lateralization in the nervous system*, ed. S. Harnad, R. W. Doty, L. Goldstein, J. Jaynes, and G. Krauthamer, 23–44. New York: Academic.

Nowak, M. A., J. B. Plotkin, and V.A.A. Jansen. 2000. The evolution of syntactic communication. *Nature* 404: 495–98.

Oldfield, R. C. 1971. The assessment and analysis of handedness: The Edinburgh inventory. *Neuropsychologia* 9: 97–113.

Orton, S. T. 1937. *Reading, writing and speech problems in children*. New York: Norton.

Paloutzian, R. F. and L. A. Kirkpatrick. eds. 1995. Religious influences on personal and social well-being. *Journal of Social Issues*, 51(2) (special issue).

Panksepp, J., and J. Burgdorf. 1999. Laughing rats? Playful tickling arouses high-frequency ultrasonic chirping in young rodents. In *Toward a science of consciousness III*, ed. S. Hameroff, D. Chalmers, and A. Kasniak, 231–44. Cambridge, Mass.: MIT Press.

Parr, L. A., W. D. Hopkins, and F.B.M. de Waal. 1997. Haptic discrimination in capuchin monkeys (*Cebus apella*): Evidence of manual specialization. *Neuropsychologia* 35: 143–52.

Passingham, R. E. 1982. *The human primate*. San Francisco: Freeman.

Patterson, F. 1978. Conversations with a gorilla. *National Geographic* 154: 438–65.

Patterson-Rudolph, C. 1993. *Petroglyphs and the Pueblo myths of the Rio Grande*. Albuquerque, N.M.: Avanyu.

———. 1997. *On the trail of the spider woman*. Santa Fe, N.M.: Ancient City.

Penfield, W., and E. Bouldrey. 1937. Somatic motor and sensory representation in the cerebral cortex of man as studied by electrical stimulation. *Brain* 60: 389–443.

Pepperberg, I. M. 1990. Some cognitive capacities of an African Grey parrot (*Psittacus erithacus*). *Advances in the Study of Behavior* 19: 357–409.

Petitto, L. A., and P. Marentette, 1991. Babbling in the manual mode: Evidence for the ontogeny of language. *Science* 251: 1493–96.

Pfeiffer, J. E. 1973. *The emergence of man.* London: Book Club.

———. 1985. *The emergence of humankind.* New York: Harper and Row.

Piattelli-Palmarini, M. 1980. *Language and learning: The debate between Jean Piaget and Noam Chomsky.* Cambridge, Mass.: Harvard University Press.

Pinker, S. 1994. *The language instinct.* New York: William Morrow.

———. 1997. *How the mind works.* London: Penguin Press.

———. 1999. *Words and rules.* New York: Basic Books.

Plooij, F. X. 1978. Some basic traits of language in wild chimpanzees? In *Action, gesture, and symbol*, ed. A. Lock, 111–31. London: Academic.

Poizner, H., E. Klima, and U. Bellugi. 1987. *What the hands reveal about the brain.* Cambridge, Mass.: MIT Press.

Porac, C., L. Rees, and T. Buller. 1990. Switching hands: A place for left hand use in a right hand world. In *Left-handedness: Behavioral implications and anomalies*, ed. S. Coren, 259–90. Amsterdam: Elsevier Science.

Poreh, A. M., J. Levin, H. Teves, and J. States. 1997. Mixed handedness and schizotypal personality in a non-clinical sample—the role of task demand. *Personality and Individual Differences* 23: 501–7.

Potì, P. 1997. Logical structure of young chimpanzees' spontaneous object grouping. *International Journal of Primatology* 18: 33–59.

Povinelli, D. J. 2001. *Folk physics for apes.* New York: Oxford University Press.

Povinelli, D. J., J. M. Bering, and S. Giambrone. 2000. Toward a science of other minds: Escaping the argument by analogy. *Cognitive Science* 24: 509–41.

Premack, D., and A. J. Premack. 1994. How "theory of mind" constrains language and communication. *Discussions in Neuroscience* 1: 93–105.

Provine, R. R. 2000. *Laughter: A scientific investigation.* London: Faber.

Provins, K. A. 1997. Handedness and speech: A critical appraisal of the role of genetic and environmental factors in the cerebral lateralization of function. *Psychological Review* 104: 554–71.

Pujol, J., J. Deus, J. M. Losilla, and A. Capdevila. 1999. Cerebral lateralization of language in normal left-handed people studied by functional MRI. *Neurology* 52: 1038–43.

Ramachandran, V. S., and S. Blakeslee. 1999. *Phantoms in the brain.* London: Fourth Estate.

Ramsay, S. C., L. Adams, K. Murphy, D. R. Corfield, S. Grootoonk, D. L. Bailey, R.S.J. Frackowiak, and A. Guz. 1993. Regional cerebral blood flow during volitional expiration in man—a comparison with volitional inspiration. *Journal of Physiology* 461: 85–101.

Rasmussen, T., and B. Milner. 1977. The role of early left-brain injury in determining lateralization of cerebral speech functions. *Annals of the New York Academy of Sciences* 299: 355–69.

Rayner, R. J., B. P. Moon, and J. C. Masters. 1994. The Makapansgat australopithecine environment. *Journal of Human Evolution* 24: 219–31.

Redican, W. K. 1975. Facial expression in nonhuman primates. In *Primate behavior: Developments in field and laboratory research*, ed. L. A. Rosenblum, 103–94. New York: Academic Press.

Renfrew, C. 1992. Archaeology, genetics, and linguistic diversity. *Man* 27: 445–78.

Renfrew, C., and D. Nettle. 1999. *Nostratic: Examining a linguistic macro-family.* Cambridge, U.K.: McDonald Institute for Archaeological Research.

Richards, G. 1986. Freed hands or enslaved feet? A note on the behavioral implications of ground-dwelling bipedalism. *Journal of Human Evolution* 15: 43–50.

Richman, B. 1993. On the evolution of speech: Singing as the middle term. *Current Anthropology* 34: 721–22.

Richmond, B. G., and D. S. Strait. 2000. Evidence that humans evolved from a knuckle-walking ancestor. *Nature* 404: 382–85.

Rilling, J. K., and T. R. Insel. 1999. The primate neocortex in comparative perspective using magnetic resonance imaging. *Journal of Human Evolution* 37: 191–223.

Rizzolatti, G. 1998. What happened to *Homo habilis?* (Language and mirror neurons). *Behavioral and Brain Sciences* 21: 527–28.

Rizzolatti, G., L. Fadiga, V. Gallese, and L. Fogassi. 1996. Premotor cortex and the recognition of motor actions. *Cognitive Brain Research* 3: 131–41.

Rizzolatti, G., L. Fadiga, M. Matelli, V. Bettinardi, E. Paulesu, D. Perani, and F. Fazio. 1996. Localization of grasp representation in humans by PET: Observation versus execution. *Experimental Brain Research* 111: 246–52.

Rogers, L. J. 1980. Lateralization in the avian brain. *Bird Behavior* 2: 1–12.

———. 2000. Evolution of side biases: Motor versus sensory lateralization. In *Side bias: A neuropsychological perspective*, ed. M. K. Mandal, M. B. Bulman-Fleming, and G. Tiwara, 3–4. Dortrecht: Kluver.

Ronen, A. 1998. Domestic fire as evidence for language. In *Neandertals and modern humans in Western Asia*, ed. T. Akazawa, K. Aoki, and O. Bar-Yosef. New York: Plenum.

Rousseau, J. J. 1994. *Discours sur l'origine et les fondements de l'inégalité parmi les hommes*. Orig. pub. 1775. In *Oeuvres complètes*, ed. B. Gagnebin and M. Raymond, vol. 3. Paris: Gallimard.

Ruhlen, M. 1994. *The origin of language*. New York: Wiley.

Sacks, O. 1991. *Seeing voices*. London: Picador.

Saki [H. H. Munro]. 1930. *The short stories of Saki*. Letchworth, Herts. U.K.: Garden City.

Sampson, G. 1997. *Educating Eve: The 'Language Instinct' debate*. London: Cassell.

Savage-Rumbaugh, E. S. 1986. *Ape language*. New York: Columbia University Press.

Savage-Rumbaugh, E. S., K. McDonald, R. A. Sevcik, W. D. Hopkins, and E. Rubert. 1986. Spontaneous symbol acquisition and communicative use by pygmy chimpanzees (*Pan paniscus*). *Journal of Experimental Psychology: General* 115: 211–35.

Savage-Rumbaugh, S. and R. Lewin. 1994. *Kanzi: The ape at the brink of the human mind*. New York: Wiley.

Savage-Rumbaugh, S., S. G. Shanker, and T. J. Taylor. 1998. *Apes, language, and the human mind*. New York: Oxford University Press.

Schaller, G. 1963. *The mountain gorilla*. Chicago: University of Chicago Press.

Schluter, N. D., M. Krams, M.F.S. Rushworth, and R. E. Passingham. 2001. Cerebral dominance for action in the human brain. *Neuropsychologia* 39: 105–13.

Scully, G. W. 1997. Murder by the state. *New York Times*, sec. 4, 14 December, 7. See also www.ncpa.org.

Seielstad, M. T., E. Minch, and L. L. Cavalli-Sforza. 1998. Genetic evidence for a higher female migration rate in humans. *Nature Genetics* 20: 278–80.

Sekiyama, K., S. Miyauchi, T. Imaruoka, H. Egusa, and T. Tashiro. 2000. Body image as a visuomotor transformation device revealed in adaptation to reversed vision. *Nature* 407: 374–77.

Semaw, S., P. Renne, J.W.K. Harris, C. S. Feibel, R. L. Bernor, N. Fesseha, and K. Mowbray. 1997. 2.5-million-year-old stone tools from Gona, Ethiopia. *Nature* 385: 333–36.

Semino, O., G. Passarino, P. J. Oefner, A. A. Lin, S. Arbuzova, L. E. Beckman, G. De Benedictus, P. Francalacci, A. Kouvatsi, S. Limborska, M. Marcikiae, A. Mika, B. Mika, D. Primorac, A. S. Santachiara-Benerecetti, L. L. Cavalli-Sforza, and P. A. Underhill. 2000. The genetic legacy of Paleolithic *Homo sapiens* in extant Europeans: A Y chromosome perspective. *Science* 290: 1155–59.

Senut, B., M. Pickford, D. Gommery, P. Mein, K. Cheboi, and Y. Coppens. 2001. First hominid from the Miocene (Lukeino Formation, Kenya). *Comtes Rendues de l'Académie des Sciences, Sciences de la Terre et des Planètes* 332: 137–44.

Sergent, J., E. Zuck, S. Terriah, and B. MacDonald. 1992. Distributed neural network underlying musical sight-reading and keyboard performance. *Science* 257: 106–9.

Shan-Ming, Y., P. Flor-Henry, C. Dayi, Li Tiang, Qui Shuguang, and Ma Xenxiang. 1985. Imbalance in hemisphere function in the major psychoses: A study of handedness in the People's Republic of China. *Biological Psychiatry* 20: 906–17.

Shepard, R. N. 1978. The mental image. *American Psychologist* 33: 125–37.

Shevoroshkin, V. 1990. The mother tongue: How linguists have reconstructed the ancestor of all living languages. *The Sciences*, May/June, 20–27.

Shimizu, A., M. Endo, N. Yamaguchi, H. Torii, and K. Isaka. 1985. Hand preference in schizophrenics and hand conversion in their childhood. *Acta Psychiatria* 72: 259–65.

Soffer, O, J. M. Adovasio, J. S. Illingworth, H. A. Amirkhanov, N. D. Praslov, and M. Street. 2000. Palaeolithic perishables made permanent. *Antiquity* 74: 812–21.

Stokoe, W. C. 1960. *Sign language structure.* Silver Spring, Md.: Linstok Press.

———. 1987. Sign language and the monastic use of gestures. In *Monastic sign languages*, ed. J. Uniker-Sebeok and T. A. Sebeok, 323–38). Berlin: Mouton de Gruyter.

Stokoe, W. C., D. C. Casterline, and C. G. Croneberg. 1965. *A dictionary of American Sign Language on linguistic principles.* Silver Spring, Md.: Linstok.

Stone, V. E., S. Baron-Cohen, and R. T. Knight. 1999. Frontal lobe contributions to theory of mind. *Journal of Cognitive Neuroscience* 10: 640–56.

Stringer, C. B., and P. Andrews. 1988. Genetic and fossil evidence for the origin of modern humans. *Science* 239: 1263–68.

Suddendorf, T. 1999. The rise of the metamind. In *The descent of mind*, ed. M. C. Corballis and S.E.G. Lea, 218–60. Oxford: Oxford University Press.

Suddendorf, T., and M. C. Corballis. 1997. Mental time travel and the evolution of the human mind. *Genetic, Social, and General Psychology Monographs* 123: 133–67.

Supalla, T., and R. Webb. 1995. The grammar of international sign. In *Language, gesture, and space*, ed. K. Emmorey and K. Reilly, 333–51. Hillsdale, N.J.: Erlbaum.

Swisher, C. C., III, G. H. Curtis, A. C. Jacob, A. G. Getty, A. Suprojo, and Widiasmoro. 1994. Age of the earliest known hominids in Java, Indonesia. *Science* 263: 1118–21.

Swisher, C. C., III, W. J. Rink, H. P. Anton, H. P. Schwarcz, G. H. Curtis, A. Suprijo, and Widiasmoro. 1996. Latest *Homo erectus* of Java: Potential contemporaneity with *Homo sapiens* in Southeast Asia. *Science* 274: 1870–74.

Takahat, N., S.-H. Lee, and Y. Satta. 2001. Testing multiregionality of modern human origins. *Molecular Biological Evolution* 18: 172–83.

Tanner, J. E., and R. W. Byrne. 1996. Representation of action through iconic gesture in a captive lowland gorilla. *Current Anthropology* 37: 162–73.

Tattersall, I. 1997. Out of Africa again . . . and again? *Scientific American* 276 (4): 60–70.

Taub, E., I. A. Goldberg, and P. Taub. 1975. Deafferentation in monkeys: Pointing at a target without visual feedback. *Experimental Neurology* 46: 178–86.

Thatcher, R. W., R. A. Walker, and S. Guidice. 1987. Human cerebral hemispheres develop at different rates and ages. *Science* 236: 1110–13.

Thieme, H. 1997. Lower Palaeolithic hunting spears from Germany. *Nature* 385: 807–10.

Thomas, K. 1984. *Man and the natural world*. Harmondsworth, U.K.: Penguin.

Thomson, K. S. 2000. Huxley, Wilberforce and the Oxford Museum. *American Scientist* 88 (3): 210–13.

Thoreson, C. E. 1999. Spirituality and health: Is there a relationship? *Journal of Health Psychology* 4: 291–300.

Tierney, P. 2001. *Darkness in El Dorado: How scientists and journalists devastated the Amazon*. New York: Norton.

Tobias, P. V. 1987. The brain of *Homo habilis*: A new level of organization in cerebral evolution. *Journal of Human Evolution* 16: 741–61.

———. 1998. Water and human evolution. *Out There* 35: 38–44.

Tomasello, M. 1996. Do apes ape? In *Social learning in animals: The roots of culture*, ed. J. Galef and C. Heyes, 319–46. New York: Academic.

———. 1999. *The cultural origins of human cognition*. Cambridge, Mass.: Harvard University Press.

————. 2000. Culture and cognitive development. *Current Directions in Psychological Science* 9: 37–40.

Tomasello, M., and J. Call. 1997. *Primate cognition.* New York: Oxford University Press.

Tomasello, M., and L. Camaione, 1997. A comparison of the gestural communication of apes and human children. *Human Development* 40: 7–24.

Tomasello, M., J. Call, J. Warren, G. T. Frost, M. Carpenter, and K. Nagell. 1997. The ontogeny of chimpanzee gestural signals: A comparison across groups and generations. *Evolution of Communication* 1: 223–59.

Tomasello, M., B. Hare, and B. Agnetta. 1999. Chimpanzees (*Pan troglodytes*) follow gaze direction geometrically. *Animal Behaviour* 58: 769–77.

Tooby, J., and L. Cosmides. 1989. Evolutionary psychology and the generation of culture, part 1. *Ethology and Sociobiology* 10: 29–49.

Tooby, J., and I. DeVore. 1987. The reconstruction of hominid behavioral evolution through strategic modelling. In *The evolution of human behavior: Primate models,* ed. W. G. Kinzey, 183–237. New York: State University of New York Press.

Toth, N., K. D. Schick, E. S. Savage-Rumbaugh, R. A. Sevcik, and D. M. Rumbaugh. 1993. Pan the tool-maker: Investigations into the stone toolmaking and tool-using capabilities of a bonobo (*Pan paniscus*). *Journal of Archeological Science* 20: 81–91.

Travis, L. E. 1931. *Speech pathology.* New York: Appleton.

Tyack, P. L. 2000. Dolphins whistle a signature tune. *Science* 289: 1310–11.

Underhill, P.A., P. D. Shen, A. A. Lin, L. Jin, G. Passarino, W. H. Yang, E. Kauffman, B. Bonne-Tamir, J. Bertranpetit, P. Francalacci, M. Ibrahim, T. Jenkins, J. R. Kidd, S. Q. Mehdi, M. T. Seielstad, R. S. Wells, A. Piazza, R. W. Davis, M. W. Feldman, L. L. Cavalli-Sforza, and P. J. Oefner. 2000. Y chromosome sequence variation and the history of human populations. *Nature Genetics* 26: 358–61.

Vallortigara, G., L. J. Rogers, and A. Bisazza. 1999. Possible evolutionary origins of cognitive brain function. *Brain Research Reviews* 30: 164–75.

Vaneechoutte, M., and J. R. Skoyles. 1998. The memetic origin of language: Modern humans as musical primates. *Journal of Memetics* 2: 84–117.

Volterra, V., and J. M. Iverson. 1995. When do modality factors affect the course of language acquisition? In *Language, gesture, and space,* ed. K. Emmorey and K. Reilly, 371–90. Hillsdale, N.J.: Erlbaum.

Vygotsky, L. S. 1962. *Thought and language*. Cambridge, Mass.: MIT Press.

Wada, J. A., R. Clarke, and A. Hamm. 1975. Cerebral hemispheric asymmetry in humans. *Archives of Neurology* 32: 239–46.

Walter, R. C., R. T. Buffler, J. H. Bruggemann, M.M.M. Guillaume, S. M. Berhe, B. Negassi, Y. Libsekal, H. Cheng, R. L. Edwards, R. D. von Cosel, D. Neraudeau, and M. Gagnon. 2000. Early human occupation of the Red Sea coast of Eritrea during the last interglacial. *Nature* 405: 65–69.

Ward, R., and C. Stringer. 1997. A molecular handle on the Neanderthals. *Nature* 388: 225–26.

Warren, R. M. 2000. Phonemic organization does not occur: Hence no feedback. *Behavioral and Brain Sciences* 23: 350–51.

Warrington, E. K., and R.T.C. Pratt. 1973. Language laterality in left handers assessed by unilateral ECT. *Neuropsychologia* 11: 423–28.

Wason, P. 1966. Reasoning. In *New horizons in psychology*, ed. B. M. Foss, 135–51. London: Penguin.

Watson, J. D., and F.H.C. Crick. 1953. A structure for deoxyribose nucleic acid. *Nature* 171: 737–38.

Watson, R. T., E. Valenstein, A. Day, and K. M. Heilman. 1994. Posterior neocortical systems subserving awareness and neglect. *Archives of Neurology* 51: 1014–21.

Westergaard, G. C. 1998. What capuchin monkeys can tell us about the origins of hominid material culture. *Journal of Material Culture* 3: 5–19.

Westergaard, G. C., C. Liv, M. K. Haynie, and S. J. Suomi. 2000. A comparative study of aimed throwing by monkeys and humans. *Neuropsychologia* 38: 1511–17.

Wheeler, M. A., D. T. Stuss, and E. Tulving. 1997. Toward a theory of episodic memory: The frontal lobes and autonoetic consciousness. *Psychological Bulletin* 121: 331–54.

White, T. D., G. Suwa, and B. Asfaw. 1994. *Australopithecus ramidus*, a new species of early hominid from Aramis, Ethiopia. *Nature* 371: 306–12.

———. 1995. Corrigendum to '*Australopithecus ramidus*, a new species of early hominid from Aramis, Ethiopia.' *Nature* 375: 88.

Whiten, A., J. Goodall, W. C. McGrew, T. Nishida, V. Reynolds, Y. Sugiyama, C.E.G. Tutin, R. W. Wrangham, and C. Boesch. 1999. Cultures in chimpanzees. *Nature* 399: 682–85.

Wilkins, W. K., and J. Wakefield. 1995. Brain evolution and neurolinguistic preconditions. *Behavioral and Brain Sciences* 18: 161–226.

Williams, B. 1978. *Descartes: The project of pure enquiry*. Hassocks, U.K.: Harvester.

Winkworth, A. L., P. J. Davis, R. D. Adams, and E. Ellis. 1995. Breathing patterns during spontaneous speech. *Journal of Speech and Hearing Research* 38: 124–44.

Witelson, S. F., D. L. Kigar, and T. Harvey. 1999. The exceptional brain of Albert Einstein. *The Lancet* 353: 2149–53.

WoldeGabriel, G., T. D. White, G. Suwa, P. Renne, J. Deheinzelin, W. K. Hart, and G. Heiken. 1994. Ecological and temporal placement of early Pliocene hominids at Aramis. *Nature* 371: 330–33.

Wolpoff, M. H., J. Hawks, D. W. Frayer, and K. Hunley. 2001. Modern human ancestry at the peripheries: A test of the replacement theory. *Science* 291: 293–97.

Wood, B. A. 1992. Evolution of australopithecines. In *The Cambridge encyclopedia of human evolution*, ed. S. Jones, R. Martin, and D. Pilbeam, 231–40. Cambridge: Cambridge University Press.

Wood, B., and M. Collard. 1999. The human genus. *Science* 284: 65–71.

Wundt, W. M. 1921. *Elements of folk psychology.* Trans. E. L. Schaub. New York: Macmillan. Orig. pub. 1916.

Yamei, H., R. Potts, Y. Baoyin, G. Zhengtang, A. Deino, W. Wei, J. Clark, X. Guangmao, and H. Weiwen. 2000. Mid-Pleistocene Acheulian-like stone technology of the Bose Basin, South China. *Science* 287: 1622–25.

Yates, F. A. 1966. *The art of memory.* Chicago: University of Chicago Press.

Yellen, J. E., A. S. Brooks, E. Cornelissen, M. J. Mehlman, and K. Stewart. 1995. A Middle Stone Age worked bone industry from Katanda, Upper Semliki Valley, Zaire. *Science* 268: 553–56.

Yeni-Komshian, G. H., and D. A. Benson. 1976. Anatomical study of cerebral asymmetry in the temporal lobe of humans, chimpanzees, and rhesus monkeys. *Science* 192: 387–89.

Zaidel, E. 1983. A response to Gazzaniga: Language in the right hemisphere, convergent perspectives. *American Psychologist* 38: 542–46.

Zangwill, O. L. 1976. Thought and the brain. *British Journal of Psychology* 67: 301–14.

Zipf, G. K. 1935. *The psycho-biology of language: An introduction to dynamic philology.* Boston: Houghton Mifflin.

■ SUBJECT INDEX ■

DATE DUE			
MAR 31 04			
HIGHSMITH #45114			